石油科技英语系列教材

丛书主编 ◎ 江淑娟 吴松林

Petroleum Refining

石油炼制

隋岩 秦艳霞 ◎ 编

石油工业出版社

内 容 提 要

本书是讲述石油炼制相关知识的英语学习教程,共6章,主要包括美国炼油工业发展、石油炼制过程、一次加工、二次加工、石油炼制产品和化工产品以及石油炼制与环境保护等内容。

本书可作为石油相关专业的师生和石油系统员工学习石油科技英语的参考书。

图书在版编目(CIP)数据

石油炼制/隋岩,秦艳霞编.
北京:石油工业出版社,2014.8
(石油科技英语系列教程)
ISBN 978 - 7 - 5183 - 0215 - 4

Ⅰ. 石…
Ⅱ. ①隋…②秦…
Ⅲ. 石油炼制 - 英语 - 教材
Ⅳ. H31

中国版本图书馆 CIP 数据核字(2014)第 132459 号

出版发行:石油工业出版社
　　　　(北京安定门外安华里2区1号　100011)
　　　　网　　址:http://pip.cnpc.com.cn
　　　　编辑部:(010)64523574　发行部:(010)64523620
经　　销:全国新华书店
印　　刷:北京中石油彩色印刷有限责任公司

2014 年 8 月第 1 版　2014 年 8 月第 1 次印刷
787×1092 毫米　开本:1/16　印张:19.25
字数:350 千字

定价:50.00 元
(如出现印装质量问题,我社发行部负责调换)
版权所有,翻印必究

《石油科技英语系列教程》
编 委 会

丛书主编:江淑娟　吴松林

编写人员:(按姓氏笔画排序)

马焕喜	王　红	王　琪	王　超	王化雪
王利国	历冬风	白雪晴	乔洪亮	朱年红
刘　葆	刘立安	刘亚丽	刘江红	刘杰秀
刘典忠	刘佰明	江淑娟	安丽娜	闫柯菲
孙彩光	苏亚杰	李　冰	李　健	李　超
李　滨	吴松林	谷秋菊	张　曦	张雪梅
陈　默	范金宏	卓文娟	赵　巍	姚坤明
秦艳霞	郭玉海	唐俊莉	徐国伟	高苏彤
康国强	隋　岩	韩福勇	葛多虹	董金玉
惠良虹				

前　言

　　全球石油资源分布、生产及消费三者之间存在着严重的地区失衡,中东和亚太是失衡最严重的地区,中东地区严重供过于求,亚太地区严重供不应求。因此能源行业出现了全球化发展趋势,能源国际间的交流与合作日益密切。为保证中国能源安全,中国石油和石化行业国际化与本土化发展势在必行。中国油气企业正在积极进行海外业务拓展,了解资源地区的文化背景、经济发展状况、能源开发政策并掌握其石油地质结构、油气成藏条件、开发和炼制技术等将有利于我们对资源地区的油气开发和炼制,更有力地支持中国经济的快速发展。

　　自 1993 年起,为了解决石油院校和石油职工专业英语教材的严重匮乏问题,本丛书主编陆续在黑龙江人民出版社、黑龙江科技出版社、石油工业出版社等出版了系列专业石油英语教科书,积累了一定的编写经验、培训经验和图书项目导向经验。20 年过去了,石油行业也发生了巨大的变化,新油气资源不断发现,开采与炼制等技术不断更新,海外合作区域也不断拓宽。为了适应新形势,我们经过不懈努力通过了中国石油天然气集团公司图书出版立项,开始编写一套更大规模的《石油科技英语系列教程》,既包括石油上、中、下游生产技术,也包括世界主要石油资源国的经济、贸易和文化等,目的是为读者奠定通向世界石油领域话语权的语言基础。

　　我们深感责任重大,从中国石油大学、东北大学、东北石油大学、西安石油大学及各油田石油地质研究院、设计院等单位聘请有关专家学者,确定编写体例,搜集资料。在选材上注重内容的系统性,争取覆盖本领域主要内容;语言方面,注意遴选突出科技英语语言特点的语段和篇章,并对语言使用方法做详尽解释,以英语基础知识和基本技能的培养为主。为降低学习难度,为每篇课文还配写了汉语译文,以提高学生的石油科技英语阅读、翻译及写作能力。

　　《石油炼制》分册共分为六章,第一章对美国石油炼制业的历史和发展做了简要介绍,因为美国石油炼制技术居世界前列;第二、第三和第四章对石油粗炼和精炼等系列炼制过程做了详细描述;第五章着重介绍了各类石油炼制产品,如汽油、煤油、柴油、燃料油、润滑油、石油蜡和沥青等;第六章对石油炼制和环境保护问题进行了探讨。为了让学习者更好地理解课文内容,对每一篇课文均给出导语,列出重点专业词汇及词组,对重难点句子进行讲解,并跟进与重点内容相关的问题,使读者在回答问题的同时可以巩固对课文的理解,进而掌握相关专业知识。

《石油炼制》分册第一章至第三章由中国石油大学(北京)的隋岩编写,第四章至第六章由西安石油大学的秦艳霞编写。中国石油大学(北京)的王晓阳、肖家栋和林丹帮助做了大量的资料查找和整理工作,在此一并表示感谢。

由于作者时间和水平有限,且书中内容涉及面较广,可供参考、借鉴的资料不多,难免出现不尽如人意的地方,敬请专家和读者批评指正。

丛书主编:江淑娟　吴松林

2014 年 3 月

Content

Chapter 1　American Refinery Industry Development

1.1　Petroleum prior to the Modern Era

⌸ Guidance to Reading

Due to a very limited utility, petroleum had a small market prior to the industrial era. Oil's age of illumination sparked the modern petroleum industry. Rockefeller's horizontal combination in the refining sector initiated competition and resulted in monopoly of the industry. And Standard Oil was forced to dissolution...

⌸ Text

Prior to the industrial era, petroleum had a very limited utility. Records show human uses for oil in the form of natural **seepage** of **asphaltic bitumen** as far back as ancient Mesopotamia prior to 3000 BC. Employed as **mastic** in construction, a waterproofing for ships, and a medicinal **poultice**, this oil apparently did have a small market. Later, Near Eastern cultures successfully **distilled** natural crude oil to obtain lamp fuel and the basic ingredient for "Greek fire", a military **incendiary** introduced with great effect by the Byzantine Empire against the rigging of attacking ships. In the West, limited supplies of crude oil seepage were confined mostly to medicinal use. In a parallel development, a small Russian industry developed around the presence of natural seepage in the Baku region, initially also for medicinal use. However, it took a 19th – century shortage of lamp oil, precipitated by a whale oil crisis and relative scarcity of coal oil, to spark the modern petroleum industry.

Oil's Age of Illumination

In the 1850s, experiments with **refined** petroleum demonstrated that it could serve as a satisfactory lamp oil (in **kerosene** form). The problem was that it was not available in suitable quantities. On August 27, 1859, Drake's **rig** struck oil at 69 ft near Oil Creek, a short distance from the town of Titusville,

Pennsylvania. The oil age was born. The market for illuminating oil was **insatiable** and the growth of the industry was sudden and dramatic. Crude oil production increased from only 2000 barrels in 1859 to 4. 8 million barrels a decade later and 5,350,000 barrels in 1871. Between the 1860s and 1900, crude oil and oil products exported ranged from one-third to three-fourths of total U. S. production.

Ease of entry defined all aspects of the business, and capital investment was minimal. Refining technology was primitive, consisting of heating crude in a **still** in order to obtain the desired kerosene fraction; other constituent fractions such as **gasoline** were often run into streams or onto the ground in ditches. The evolution of the legal doctrine of law of capture by the Pennsylvania courts also contributed to wasteful practices in the oil fields. Several different landowners or leaseholders might sit astride an underground **oil pool** . Drawing an **analogy** with game being legally captured if it were lured onto another's land, and citing precedent in English common law, the courts upheld the right of the property owner to capture oil that had **migrated** under the owner's land. Limited understanding of oil geology at that time mistakenly led to the belief that oil flowed in underground rivers. Each owner or leaseholder rushed to drill and **pump** before their neighbors **depleted** the oil pool. Oil markets became characterized by alternate periods of oil **glut** followed quickly by scarcity. This wasteful practice continued as the oil frontier moved westward from Pennsylvania, Ohio, and Indiana in the latter 19th century to Texas, California, and the midcontinent in the 20th century.

Competition and Monopoly

The young John D. Rockefeller entered this rough-and-tumble business in 1863. He had cut his teeth as a partner in the wholesale grocery business in Cleveland, doing well as **purveyor** of supplies to Ohio military units. However, Cleveland possessed several locational advantages that encouraged its development as an important refining center. Located on Lake Erie, close to the Pennsylvania oil fields, it had railroad trunkline connections to major eastern markets, a readily available workforce, and sources of financial capital. The industry, then as today, consisted of four main functional sectors: production, refining, transportation, and **marketing** . Production entails all the activities involved in "getting it out of the ground" . This includes **exploration** , drilling,

leasing ,pumping, **primary and secondary recovery** techniques, and all other associated activities. Refining entails all relevant processes developed to obtain usable products from crude oil in its natural state. Initially, this included only **fractional distillation** ,the heating of crude to boil off its constituent parts, but has evolved today to include a host of sophisticated specialty practices. Transportation, the third element of the industry, **encompasses** the movement of crude oil from the oil field to the refinery and the shipment of refined product to market. In the beginning of the industry in the 1860s and 1870s, this meant by barge, railroad car, or horse – drawn wagon. Oil transport evolved to include petroleum pipelines, tanker trucks on the highway, and fleets of tankers on the high seas. Marketing envelops the distribution and sale of petroleum products to the consumer.

Rockefeller's early endeavors were concentrated in the refining sector of the industry, achieving growth through horizontal combination; he did not adopt a strategy of vertical integration until later. First as a partnership, then organized as a corporation in 1870, Rockefeller and associates' Standard Oil Company proceeded to achieve what was later termed the conquest of Cleveland. Using tactics that one could define as either **shrewd** or unethical, Rockefeller succeeded in dominating the Cleveland refining business. Concentrating on transportation as the key to control, Standard used its size to obtain preferential shipping arrangements with the railroads through the use of rebates and then sought to eliminate competition with a series of pooling and cartel agreements. Rockefeller's attempts to control the refining business with the South Improvement Company and later the National Refiners Association remain textbook examples of business strategy in the late 19th century. When Rockefeller found informal agreements inadequate to achieve tight industry control, he changed his approach to one of merger and acquisition.

By 1878, Standard controlled more than 90% of total U. S. refining capacity. To formalize this economic empire, Rockefeller and associates created the Standard Oil Trust Agreement in 1882, later reorganized as the Standard Oil Company (New Jersey) as a legal holding company under New Jersey law in 1899. Rockefeller integrated backward in the 1880s by acquiring producing properties in Ohio, into transportation with his own pipelines, and forward into marketing with his branded cans of kerosene for domestic sale and export. The

Standard near monopoly became the hated enemy of independent oil producers, who found their prices driven down by Rockefeller buying power, and by smaller operators, who had been forced to sell out in the face of price – cutting attacks.

Standard had been under legal attack in state courts by its enemies and competitors for years, but the passage of the Sherman Antitrust Act in 1890 had lain the groundwork for federal challenge. With a change in political climate during the presidency of Theodore Roosevelt toward progressivism, federal investigations went forward, **culminating** in the forced **dissolution** of Standard Oil into several constituent companies in 1911. This did not hurt Rockefeller personally too much because he remained a stockholder in these many companies, and historians have argued that the former monopoly had simply transformed into an **oligopoly** of large vertically integrated companies.

🔲 Words and Expressions

seepage	n.	渗流
mastic	n.	泥胶
poultice	n.	膏药
distill	v.	蒸馏
incendiary	n.	燃烧弹
illumination	n.	照明
refine	v.	炼制
kerosene	n.	煤油
rig	n.	钻机
insatiable	adj.	不能满足的
still	n.	蒸馏器
gasoline	n.	汽油
analogy	n.	类比
migrate	v.	运移
pump	v.	用泵抽
deplete	v.	耗尽
glut	v.	供给过剩
monopoly	n.	垄断
purveyor	n.	供应商

marketing	n. 销售
exploration	n. 勘探
lease	v. 租赁
encompass	v. 包含,包围
shrewd	adj. 精明的
culminate	v. 导致
dissolution	n. 破裂,解散
oligopoly	n. 寡头

Technical Terms

asphaltic bitumen	沥青
oil pool	油田
primary recovery	一次采油
secondary recovery	二次采油
fractional distillation	分馏

Language Focus

1. Employed as mastic in construction, a waterproofing for ships, and a medicinal poultice, this oil apparently did have a small market.

(参考译文为:这种柏油沥青可以被用作建筑泥胶、船舶防水以及医疗膏药,很显然它的市场很小。)

本句中"employ"一词意为"利用",在这里用作过去分词作状语,说明动词与主语"the oil"为被动关系,意为"柏油沥青可以被用于……"。

2. However, it took a 19th – century shortage of lamp oil, precipitated by a whale oil crisis and relative scarcity of coal oil, to spark the modern petroleum industry.

本句的主体结构为"it takes + n. + n. + to - inf","take"在此处作需要、要求讲,即做某事需要某些条件,"precipitated by a whale oil crisis and relative scarcity of coal oil"部分为插入成分,做"19th – century shortage of lamp oil"的定语。

3. The problem was that it was not available in suitable quantities.

此句中"that"位于"was"之后,引导表语从句,"it was not available in suitable quantities"解释说明前面提到的"problem"。

4. Refining technology was primitive, consisting of heating crude in a still in order to obtain the desired kerosene fraction; other constituent fractions such as gasoline were often run into streams or onto the ground in ditches.

此句中间用分号隔开，表明是意思相对独立的两部分，比之连词"and"，分号更为醒目，表达效果更直接。前后两部分主谓结构不同，但内容相关，因而整合为一个长句。

5. Located on Lake Erie, close to the Pennsylvania oil fields, it had railroad trunkline connections to major eastern markets, a readily available workforce, and sources of financial capital.

此句主语为"it"，谓语动词为"had located"和"close"两个形容词分别修饰主语"it"。句子后半部分"workforce"和"sources"为并列成分，进一步解释说明"eastern markets"。

6. The Standard near monopoly became the hated enemy of independent oil producers, who found their prices driven down by Rockefeller buying power, and by smaller operators, who had been forced to sell out in the face of price - cutting attacks.

此句中第一个"who"引导定语从句，修饰"oil producer"；句中两处"by"为并列成分，列举了拉低"independent oil producers"价格的原因；最后一个"who"引导的定语从句修饰 smaller operators。

🔲 Reinforced Learning

Ⅰ. Answer the following questions for a comprehension of the text.

1. What sparked the modern petroleum industry?

2. Did the refined petroleum be used as a satisfactory lamp oil?

3. What are the advantages for Cleveland as an important refining center?

4. What are the main functional sectors in oil industry? And explain them in detail.

5. Was Rockefeller hurt after the dissolution of Standard Oil and why?

Ⅱ. Multiple choice: choose the correct one from the alternative answers to give the exact meaning of the word underlined.

1. Prior to the industrial era, petroleum had a very limited utility.

A. function   B. effect   C. usage   D. service

2. Employed as mastic in construction, a waterproofing for ships, and a

medicinal poultice, this oil apparently did have a small market.

   A. used   B. occupied   C. worked   D. sold

   3. However, it took a 19th – century shortage of lamp oil, precipitated by a whale oil crisis and relative scarcity of coal oil, to spark the modern petroleum industry.

   A. caused   B. worsen   C. participated   D. partaken

   4. The problem was that it was not available in suitable quantities.

   A. fit   B. satisfactory   C. right   D. proper

   5. Refining technology was primitive, consisting of heating crude in a still in order to obtain the desired kerosene fraction.

   A. advanced   B. wanted   C. designed   D. processed

   6. This wasteful practice continued as the oil frontier moved westward from Pennsylvania, Ohio, and Indiana in the latter 19th century to Texas, California, and the midcontinent in the 20th century.

   A. exercise   B. behavior   C. habit   D. procedure

   7. Initially, this included only fractional distillation, the heating of crude to boil off its constituent parts, but has evolved today to include a host of sophisticated specialty practices.

   A. something distinctive   B. something particular

   C. something special   D. something professional

   8. Marketing envelops the distribution and sale of petroleum products to the consumer.

   A. mails   B. includes   C. encompasses   D. surrounds

   9. Standard used its size to obtain preferential shipping arrangements with the railroads through the use of rebates and then sought to eliminate competition with a series of pooling and cartel agreements.

   A. largeness   B. space   C. extent   D. amount

   10. When Rockefeller found informal agreements inadequate to achieve tight industry control, he changed his approach to one of merger and acquisition.

   A. unsatisfactory   B. insufficient   C. incompetent   D. unnecessary

   III. Multiple choice: read the four suggested translations and choose the best answer.

   1. In the 1850s, experiments with refined petroleum demonstrated that it

could serve as a <u>satisfactory</u> lamp oil (in kerosene form).

    A. 恰当的   B. 令人满意的 C. 符合要求的 D. 如愿的

    2. The evolution of the legal doctrine of law of capture by the Pennsylvania courts also <u>contributed</u> to wasteful practices in the oil fields.

    A. 有助于   B. 导致

    C. 为……做贡献   D. 有利于

    3. Several different landowners or leaseholders might <u>sit astride</u> an underground oil pool.

    A. 两腿分开站着 B. 独自占有   C. 分享   D. 控制

    4. Standard used its size to obtain <u>preferential</u> shipping arrangements with the railroads through the use of rebates and then sought to eliminate competition with a series of pooling and cartel agreements.

    A. 优惠的   B. 优先的   C. 有利的   D. 有益的

    5. The Standard near monopoly became the hated enemy of independent oil producers, who found their prices driven down by Rockefeller <u>buying power</u>, and by smaller operators, who had been forced to sell out in the face of price – cutting attacks.

    A. 购买力   B. 购买力量   C. 购买力度   D. 购买额度

IV. Put the following sentences into Chinese.

    1. The problem was that it was not available in suitable quantities.

    2. Oil transport evolved to include petroleum pipelines, tanker trucks on the highway, and fleets of tankers on the high seas.

    3. Ease of entry defined all aspects of the business, and capital investment was minimal.

    4. Oil markets became characterized by alternate periods of oil glut followed quickly by scarcity.

    5. Rockefeller's attempts to control the refining business with the South Improvement Company and later the National Refiners Association remain textbook examples of business strategy in the late 19th century.

V. Put the following paragraphs into Chinese.

    1. The market for illuminating oil was insatiable and the growth of the industry was sudden and dramatic. Crude oil production increased from only 2000 barrels in 1859 to 4.8 million barrels a decade later and 5,350,000 barrels in

1871. Between the 1860s and 1900, crude oil and oil products exported ranged from one – third to three – fourths of total U. S. production.

2. Standard had been under legal attack in state courts by its enemies and competitors for years, but the passage of the Sherman Antitrust Act in 1890 had lain the groundwork for federal challenge. With a change in political climate during the presidency of Theodore Roosevelt toward progressivism, federal investigations went forward, culminating in the forced dissolution of Standard Oil into several constituent companies in 1911. This did not hurt Rockefeller personally too much because he remained a stockholder in these many companies, and historians have argued that the former monopoly had simply transformed into an oligopoly of large vertically integrated companies.

1. 2　Technological Change and Energy Transition

Guidance to Reading

From cable tool drilling to rotary drilling, light cracking to thermal cracking, primary to enhanced recovery techniques, technological change has improved petroleum production. More superior quality gasoline was produced instead of kerosene to meet the demand of the increasing automobiles. What do you think of Rockefeller's strategy of consolidation as well as alternative liquid fuels at the time, necessary?

Text

If the Drake well in 1859 had trumpeted the birth of the illuminating oil industry, Spindletop marked the birth of its new age of energy. Spindletop and subsequent other new western fields provided vast oil for the growing economy but also enabled firms such as Gulf, Texas, and Sun Oil to gain a substantial foothold. The gulf region was underdeveloped and attractive to young companies, and the state of Texas was hostile toward the Standard Oil monopoly. Asphaltic – based Spindletop crude made inferior grades of kerosene and **lubricants** but yielded a satisfactory **fuel oil**. New firms, such as Gulf and Texaco, easily integrated forward into the refining, transportation, and marketing of this fuel oil in the coal – starved Southwest.

In 1900, Standard Oil (New Jersey) controlled approximately 86% of all crude oil supplies, 82% of refining capacity, and 85% of all kerosene and gasoline sold in the United States. On the eve of the 1911 court **decree** , Standard's control of crude production had declined to approximately 60% ~ 65% and refining capacity to 64%.

Moreover, Standard's competitors now supplied approximately 70% of the fuel oil, 45% of the lubricants, 33% of **gasoline** and **waxes** , and 25% of the kerosene in the domestic market. Newly formed post – Spindletop companies such as Gulf and Texaco, along with **invigorated** older independent firms such as Sun and Pure, had captured significant market share. Meanwhile, Standard, heavily invested in the traditional kerosene business, was slow to move into the production and marketing of gasoline.

At approximately the same time, the **automobile** , which had appeared in the 1890s as a novelty, had begun to shed its elitist image. The introduction of mass – produced and relatively inexpensive vehicles, led by the 1908 Ford Model T, very quickly influenced developments in the oil business. In 1908, the total output of the U. S. auto industry was 65,000 vehicles. In less than a decade, Ford alone sold more than 500,000 units annually. Within that same decade, both the volume of production and the total value of gasoline passed those of kerosene. Now the oil industry was becoming increasingly concerned about how it could boost production of gasoline to meet demand.

As the oil industry faced real or apparent shortages of supply, one response was technological. New production methods such as the introduction of **rotary drilling** early in the century increased crude supplies, and refinery innovations enabled crude stocks to be further extended. However, as the demand for gasoline increased, new oil discoveries in Texas, California, and Oklahoma in the early 1900s proved insufficient. Once discarded as relatively useless, this lighter fraction typically constituted 10% ~ 15% of a barrel of crude oil. Refiners stretched the gasoline fraction by including more of the heavier kerosene, but this resulted in an inferior motor fuel.

One could enrich the blend of gasoline with the addition of what would later be termed **higher octane product** obtained from selected **premium crudes** or "natural" or "casing head" gasoline yielded from highly **saturated** natural gas. The most important breakthrough, however, occurred with the introduction of **thermal cracking technology** in 1913. The industry had employed light cracking, the application of heat to distillation to literally rearrange **hydrocarbon molecules**, since the 1860s to obtain higher yields of kerosene from **feed stock** . By dramatically increasing the temperature and pressure of his cracking stills, Burton discovered that he could double the output of gasoline obtained over previous fractional distillation and cracking methods. An additional advantage was that this gasoline was of generally superior quality.

Abundance, Scarcity, and Conservation

From its earliest beginnings in the 19th century, the oil industry had to deal with the recurrent feast or famine that accompanied the alternate discovery of new fields followed by their depletion. The projections of future supply in the United States published by the U. S. Geologic Survey in 1908 predicted total depletion of U. S. **reserves** by 1927, based on then current levels of consumption, and the secretary of the interior warned President Taft in 1909 of an **impending** oil shortage. Driven by the rush to get oil out of the ground by the law of capture, the industry's production sector remained chaotic and highly unpredictable as the "forest of **derricks** " that defined each field moved ever westward. Within the context of this volatile business one can read the arguments of the Standard Oil (New Jersey) attorneys defending the company from antitrust attack in a different light. They maintained that Rockefeller's strategy of consolidation, denounced as evil monopoly, in fact represented a rational attempt to impose order and stability on an unstable industry.

A new era of **conservation** as well as the first significant enforcement of antitrust law had arrived on the political scene at the beginning of the 20th century with the presidential administration of Theodore Roosevelt. It is important to differentiate the conservation approach, best defined as wise utilization of natural resources, from the preservationist approach of Roosevelt contemporary John Muir, founder of the Sierra Club. Roosevelt and his key advisers did not want to lock away America's resources but did favor planning and responsible **exploitation** . One should interpret the efforts of the U. S. Geological Survey and the Bureau of Mines (created in 1910) in this light. These two institutions provided some attention to petroleum − related issues, but the creation of the Bureau of Mines' petroleum division in 1914 signaled a **heightened** federal government interest in oil conservation. Work centered at the division's Bartlesville, Oklahoma Petroleum Experiment Station focused on **reservoir** behavior, efficient drilling practices , **well spacing** , and secondary recovery techniques and **disseminated** knowledge to industry through influential Bureau of Mines publications.

In the 1920s , petroleum conservationists began to encourage **unitization** or the unit management of oil pools as a central alternative to the wasteful law of capture. This approach encouraged each pool to be operated as a cooperative unit , with the individual leaseholder's percentage share defining the amount of oil that could be pumped out. Operators would drill fewer wells , produce oil at a controlled rate , and either use or return to the **producing zone** under pressure all natural gas obtained along with the oil rather than the standard practice of **flaring** it or **venting** it into the atmosphere.

A perceived scarcity of petroleum reserves at the beginning of the 1920s , coupled with increased demand , stimulated other technological developments. The Bureau of Mines conducted extensive research into enhanced and secondary recovery techniques such as the **waterflooding** of older fields to boost production. Integrated firms developed their own thermal cracking technologies in efforts to circumvent the Burton patents held by Standard of Indiana and increased their own gasoline output. There was also brief **flirtation** with alternative liquid fuels in the early part of the decade. Standard Oil (New Jersey) marketed a 25% ethanol – gasoline blend in 1922 – 1923 and later obtained the basic German patents for **coal liquefaction** . There was also a boom in Western shale oil in the 1920s , which has left an interesting if eccentric history. All these liquid fuel alternatives would soon prove unnecessary as new strikes of oil again dampened anxiety about shortages.

🔲 Words and Expressions

lubricant	n.	润滑油
decree	v.	判决
gasoline	n.	汽油
wax	n.	蜡
invigorated	adj.	生机勃勃的
automobile	n.	汽车
saturated	adj.	饱和的
hydrocarbon	n.	烃,碳氢化合物
molecule	n.	分子

reserve	n. 储备,储量
impending	adj. 即将发生的
derrick	n. 井架,钻井
conservation	n. 保护,保存
exploitation	n. 开发
heighten	v. 加强,巩固
reservoir	n. 油藏
disseminate	v. 传播
unitization	n. 统一化
flare	v. 燃烧
vent	v. 排放
waterflooding	n. 水驱
flirtation	n. 一时兴起

Technical Terms

fuel oil	燃料油
rotary drilling	旋转钻探
higher octane product	高辛烷值产品
premium crude	高级原油
thermal cracking technology	热裂解技术
feed stock	原料
well spacing	布井
producing zone	产油层
coal liquefaction	煤液化

Language Focus

1. Newly formed post – Spindletop companies such as Gulf and Texaco, along with invigorated older independent firms such as Sun and Pure, had captured significant market share.

（参考译文为：在斯潘德尔托普油田发现之后成立的新公司,如 Gulf 和 Texcao,连同 Sun 和 Pure 这生机勃勃的老牌公司一起,已经占据了巨大的市场份额。）

此句主语为"newly formed post – Spindletop companies",谓语动词为"had captured",该过去分词的谓语表明这些公司已经先一步占有了市场份额。

2. The introduction of mass – produced and relatively inexpensive vehicles, led by the 1908 Ford Model T, very quickly influenced developments in the oil business.

本句句子主体框架为"the introduction...influenced...","led by the 1908 Ford Model T"为过去分词,修饰"mass – produced and relatively inexpensive vehicles",意为:以福特 T 型车为代表的批量生产价格相对低廉的汽车。

3. One could enrich the blend of gasoline with the addition of what would later be termed higher octane product obtained from selected premium crudes or "natural" or "casing head" gasoline yielded from highly saturated natural gas.

(参考译文为:一种从选定原油或高度饱和的天然气中提炼获取的辛烷被人们用以改善汽油的组成。)

不定代词"one"做主句的主语,谓语为"enrich",其后是 with 引导的介词结构。"obtained from..."为过去分词做后置定语,说明 higher octane product 的来源。句子最后的 yielded from highly saturated natural gas 做定语修饰"casing head gasoline"。"casing head" gasoline 意为"天然原油"。

4. By dramatically increasing the temperature and pressure of his cracking stills, Burton discovered that he could double the output of gasoline obtained over previous fractional distillation and cracking methods.

本句以 by 引导的方式状语开头,意为通过大幅度提高温度和压力的方法,Burton 可能会比以前增加一倍的汽油产量。句中 double 为动词,意为"使……加倍",例如 The world population is doubling every thirty – five years. 世界人口每 35 年翻一番。

5. All these liquid fuel alternatives would soon prove unnecessary as new strikes of oil again dampened anxiety about shortages.

句中"as new strikes of oil again dampened anxiety about shortages",在此句中"as"为连词引导原因状语从句,意为:因为又一批新油田的发现,削弱了人们对油荒的担忧。

Reinforced Learning

I. Answer the following questions for a comprehension of the text.

1. What role did Spindletop and other western fields play?

2. How to enrich the blend of gasoline in early days?

3. What is the most important breakthrough in oil industry in 1913?

4. What was the President Roosevelt's attitude towards resources?

5. What were the new methods used in petroleum conservation in the 1920s?

Ⅱ. Multiple choice: choose the correct one from the alternative answers to give the exact meaning of the word underlined.

1. If the Drake well in 1859 had trumpeted the birth of the illuminating oil industry, Spindletop marked the birth of its new age of energy.

A. showed   B. bellowed   C. declared   D. marked

2. The gulf region was underdeveloped and attractive to young companies, and the state of Texas was hostile toward the Standard Oil monopoly.

A. unfriendly   B. against   C. aggressive   D. unfavorable

3. New firms, such as Gulf and Texaco, easily integrated forward into the refining, transportation, and marketing of this fuel oil in the coal – starved Southwest.

A. lacking   B. hungry   C. needed   D. wanted

4. As the oil industry faced real or apparent shortages of supply, one response was technological.

A. answer   B. reaction   C. aspect   D. responsibility

5. One could enrich the blend of gasoline with the addition of what would later be termed higher octane product obtained from selected premium crudes or "natural" or "casing head" gasoline yielded from highly saturated natural gas.

A. improve the quality   B. increase

C. produce more   D. enhance

6. Within the context of this volatile business one can read the arguments of the Standard Oil (New Jersey) attorneys defending the company from antitrust attack in a different light.

A. way   B. color   C. attitude   D. word

7. A new era of conservation as well as the first significant enforcement of antitrust law had arrived on the political scene at the beginning of the 20th century with the presidential administration of Theodore Roosevelt.

A. implement   B. strengthening   C. formulation   D. proposition

8. It is important to differentiate the conservation approach, best defined as wise utilization of natural resources, from the preservationist approach of Roose-

velt contemporary John Muir, founder of the Sierra Club.

A. disturb   B. discriminate   C. distinguish   D. dispute

9. Roosevelt and his key advisers did not want to lock away America's resources but did favor planning and responsible exploitation.

A. essential   B. controlling   C. basic   D. noted

10. This approach encouraged each pool to be operated as a cooperative unit, with the individual leaseholder's percentage share defining the amount of oil that could be pumped out.

A. terming   B. explaining   C. showing   D. setting

III. Multiple choice: read the four suggested translations and choose the best answer.

1. Meanwhile, Standard, heavily invested in the traditional kerosene business, was slow to move into the production and marketing of gasoline.

A. 搬进   B. 向……靠近   C. 进军   D. 消磨时间

2. Now the oil industry was becoming increasingly concerned about how it could boost production of gasoline to meet demand.

A. 担心   B. 忧虑   C. 关注   D. 与……有关

3. Refiners stretched the gasoline fraction by including more of the heavier kerosene, but this resulted in an inferior motor fuel.

A. 归因于   B. 导致   C. 由于   D. 因为

4. The industry had employed light cracking, the application of heat to distillation to literally rearrange hydrocarbon molecules, since the 1860s to obtain higher yields of kerosene from feed stock.

A. 雇用   B. 占用   C. 利用   D. 从事

5. From its earliest beginnings in the 19th century, the oil industry had to deal with the recurrent feast or famine that accompanied the alternate discovery of new fields followed by their depletion.

A. 交易   B. 处理   C. 涉及   D. 讨论

IV. Put the following sentences into Chinese.

1. The most important breakthrough, however, occurred with the introduction of thermal cracking technology in 1913.

2. Once discarded as relatively useless, this lighter fraction typically constituted 10% ~ 15% of a barrel of crude oil.

3. From its earliest beginnings in the 19th century, the oil industry had to deal with the recurrent feast or famine that accompanied the alternate discovery of new fields followed by their depletion.

4. They maintained that Rockefeller's strategy of consolidation, denounced as evil monopoly, in fact represented a rational attempt to impose order and stability on an unstable industry.

5. Now the oil industry was becoming increasingly concerned about how it could boost production of gasoline to meet demand.

V. Put the following paragraphs into Chinese.

1. In 1900, Standard Oil (New Jersey) controlled approximately 86% of all crude oil supplies, 82% of refining capacity, and 85% of all kerosene and gasoline sold in the United States. On the eve of the 1911 court decree, Standard's control of crude production had declined to approximately 60% ~ 65% and refining capacity to 64%.

2. In the 1920s, petroleum conservationists began to encourage unitization or the unit management of oil pools as a central alternative to the wasteful law of capture. This approach encouraged each pool to be operated as a cooperative unit, with the individual leaseholder's percentage share defining the amount of oil that could be pumped out. Operators would drill fewer wells, produce oil at a controlled rate, and either use or return to the producing zone under pressure all natural gas obtained along with the oil rather than the standard practice of flaring it or venting it into the atmosphere.

1.3 The U. S. Petroleum Refining Industry

🔲 Guidance to Reading

Petroleum refining is one of the leading manufacturing industries in the United State. Most U. S. crude oil distillation capacity is owned by large, integrated companies with multiple high capacity refining facilities. Environmental and safety regulations are the most important factor affecting petroleum refining, forcing the petroleum refining industry to make substantial investments in upgrading certain refinery processes to reduce emissions and alter product compositions.

🔲 Text

Background

Petroleum refining is one of the leading manufacturing industries in the U-
nited States in terms of its share of the total value of shipments of the
U. S. economy. In relation to its economic importance, however, the industry is
comprised of relatively few companies and facilities. The number of refineries
operating in the U. S. can vary significantly depending on the information
source.

1. Product Characterization

Petroleum refining is the physical and chemical separation of crude oil into
its major **distillation fractions** which are then further processed through a series
of separation and conversion steps into finished petroleum products. The primary
products of the industry fall into three major categories: fuels (**motor gasoline**,
diesel and distillate fuel oil, **liquefied petroleum gas** , jet fuel, **residual fuel oil**,
kerosene, and coke); finished nonfuel products (solvents, **lubricating oils**,
greases, **petroleum wax**, **petroleum jelly** , asphalt, and coke); and chemical
industry feedstocks (**naphtha**, **ethane**, **propane**, **butane**, **ethylene**, propylene,
butylenes, butadiene, **benzene**, toluene, and xylene). These petroleum products
comprise about 40 percent of the total energy consumed in the U. S. A (based
on BTUs consumed) and are used as primary input to a vast number of prod-
ucts, including: **fertilizers**, **pesticides**, **paints**, waxes, **thinners**, **solvents** , cleaning fluids, **detergents**, **refrigerants**, **antifreeze**, **resins**, **sealants**, **insulations**, la-
tex, rubber compounds, hard plastics, **plastic sheeting**, **plastic foam** and **synthetic fibers** . About 90 percent of the petroleum products used in the U. S. are
fuels with motor gasoline accounting for about 43 percent of the total.

2. Industry Size

Generally, the petroleum refining industry can be characterized by a relatively small number of large facilities. The Department of Energy reported 176
operating petroleum refineries in 1994 with a total **crude oil distillation capacity** of approximately 15 million barrels per day. Most U. S. crude oil distillation
capacity is owned by large, integrated companies with multiple high capacity refining facilities. Small refineries with capacities below 50,000 barrels per day,
however, do play a significant role in the industry, making up about half of all

facilities, but only 14 percent of the total crude distillation capacity.

A relatively small number of people are employed by the petroleum refining industry in relation to its economic importance. **The Bureau of the Census** estimates that 75,000 people were directly employed by the industry in 1992. However, the industry also indirectly employs a significant number of outside contractors for many refinery operations, both routine and non – routine. The value of product shipments sold by refining **establishments** was estimated to be $136 billion in 1992. This accounts for about 4 percent of the value of shipments for the entire U. S. manufacturing sector. Based on the number of people directly employed by refineries, the industry has a high value of shipments per employee of $1.8 million. In comparison, the value of shipments per employee for the steel manufacturing industry was $245,000 for the same year. The Bureau of Census employment data for 1992 indicated that 60 percent of petroleum refineries had over 100 employees.

3. Economic Trends

The United States is a net importer of crude oil and petroleum products. In 1994, imports accounted for more than 50 percent of the crude oil used in the U. S. and about 10 percent of finished petroleum products. The imported share of crude oil is expected to increase as U. S. demand for petroleum products increases and the domestic production of crude oil declines. Imported finished petroleum products serve specific market **niches** arising from logistical considerations, regional shortages, and long – term trade relations between suppliers and refiners. Exports of refined petroleum products, which primarily consist of petroleum coke, residual fuel oil, and distillate fuel oil, account for about four percent of the U. S. refinery output. Exports of crude oil produced in the U. S. account for about one percent of the total U. S. crude oil produced and imported.

The petroleum refining industry in the U. S. has felt considerable economic pressures in the past decade arising from a number of factors, including increased costs of labor, compliance with new safety and environmental regulations and the **elimination** of government subsidies through the Crude Oil Entitlements Program which had encouraged smaller refineries to add capacity throughout the 1970s. A rationalization period began after crude oil pricing and entitlements were **decontrolled** in early 1981. The market determined that there was surplus capacity and the margins dropped to encourage the closure of the least efficient capacity. Reflecting these pressures, numerous facilities have closed in recent years. Between 1982 and 1994, the number of U. S. refineries as determined by the Department of Energy dropped from 301 to 176. Most of these closures have involved small facilities refining less than 50,000 barrels of crude oil per day. Some larger facilities, however, have also closed in response to economic pressures.

Industry representatives cited complying with the increasing environmental regulations, particularly, the requirements of the Clean Air Act Amendments of 1990, as the most important factor affecting petroleum refining in the 1990s. Despite the closing of refineries in recent years, total refinery output of finished products has remained relatively steady with slight increases in the past two years. Increases in refinery outputs are attributable to higher utilization rates of refinery capacity, and to **incremental** additions to the refining capacity at existing facilities as opposed to construction of new refineries.

Demand for refined petroleum products is expected to increase slowly with the growth of the U. S. economy. The rate of increase will average about 1.5 percent per year, which is slower than the expected growth of the economy. This slower rate of increase of demand will be due to increasing prices of petroleum products as a result of the development of substitutes for petroleum products, and rising costs of compliance with environmental and safety requirements.

Recent and future environmental and safety regulatory changes are expected to force the petroleum refining industry to make substantial investments in upgrading certain refinery processes to reduce emissions and alter product compositions. For example, industry estimates of the capital costs to comply with the 1990 Clean Air Act Amendments, which mandates specific product compositions are about $ 35 to $ 40 billion. There is concern that in some cases it may be more **economical** for some refineries to close down partially or entirely rather than upgrade facilities to meet the new standards. In fact, the U. S. Departments of Energy and Commerce expect refinery shutdowns to continue through the 1990s; however, total crude oil distillation capacity is expected to remain relatively stable as a result of increased capacity and utilization rates at existing facilities. Increases in demand for finished petroleum products will be filled by increased imports.

Words and Expressions

grease	n. 润滑脂
naphtha	n. 石脑油
ethane	n. 乙烷
propane	n. 丙烷
butane	n. 丁烷
ethylene	n. 乙烯
benzene	n. 苯
fertilizer	n. 化肥
pesticide	n. 农药
paint	n. 涂料
thinner	n. 稀释剂
solvent	n. 溶剂
detergent	n. 清洁剂
refrigerant	n. 制冷剂
antifreeze	n. 抗冻结剂
resin	n. 树脂
sealant	n. 密封剂
insulation	n. 绝缘材料
establishment	n. 企业
niche	n. (市场中)某一专门营业区
elimination	n. 消除
decontrol	v. 解除对……的管制
incremental	adj. 增长的,增值的
economical	adj. 经济的,节约的

Technical Terms

distillation fraction	馏分
motor gasoline	车用汽油
liquefied petroleum gas	液化石油气
residual fuel oil	残渣燃料油
lubricating oil	润滑油
petroleum wax	石油蜡

petroleum jelly	凡士林
petroleum jelly	凡士林
plastic sheeting	塑料布
plastic foam	塑料泡沫
synthetic fiber	合成纤维
crude oil distillation capacity	原油蒸馏产能
The Bureau of the Census	普查局

Language Focus

1. Petroleum refining is one of the leading manufacturing industries in the United States in terms of its share of the total value of shipments of the U. S. economy.

本句主体结构为"petroleum refining is..."，其中"in terms of"做状语。"Shipment"本意为"出货数量"，例如：A large shipment of coal has just arrived. 一大批煤刚刚运到，此处指代为出口量，为提喻用法。提喻指以局部代整体，例如：Many hands make light work. 人多好办事。此处以"hand"这一局部代替整体的人。

2. Small refineries with capacities below 50,000 barrels per day, however, do play a significant role in the industry, making up about half of all facilities, but only 14 percent of the total crude distillation capacity.

本句主语为"small refineries"，"with capacities below 50,000 barrels per day"部分为主语的修饰成分，"do play"为谓语。"making up about half of all facilities"为现代分词做伴随状语，且该动作与主语之间是主动关系，因而采用 – ing 形式。"do"在句中没有其他助动词时强调主动词，例如：People do in fact make mistakes. 实际上人是一定会犯错误的。

3. The petroleum refining industry in the U. S. has felt considerable economic pressures in the past decade arising from a number of factors, including increased costs of labor, compliance with new safety and environmental regulations and the elimination of government subsidies through the Crude Oil Entitlements Program which had encouraged smaller refineries to add capacity throughout the 1970s.

（参考译文为：在过去的几十年里，由于劳动力成本的增加，新安全和环保法规的颁布，以及19世纪70年代政府授权解除补贴原油贸易协定等多方面的原因，美国的大型石油炼制企业承受了巨大的经济压力，使小型炼厂不得不提高自己的炼油能力。）

本句框架为 the petroleum refining industry (in the U. S.) felt considerable pressure (in the past decade) arising from A; B; and C。

4. The market determined that there was surplus capacity and the margins dropped to encourage the closure of the least efficient capacity.

(参考译文为:由市场调节导致的生产能力过剩和生产利润下降,使许多低效工厂倒闭。)

"determine"本意为主观地决定,在此处却做"market"的谓语动词。此处作者采用拟人化手法,使市场成为支配、决定的人,使文章更为生动具体。

5. This slower rate of increase of demand will be due to increasing prices of petroleum products as a result of the development of substitutes for petroleum products, and rising costs of compliance with environmental and safety requirements.

(参考译文为:这种缓慢的增长是由于石油产品价格的提高,环保和安全要求,石油替代产品的开发,以及成本的上升等原因造成的。)

"due to"意为"由于","as a result of"意为"由于,因此",句子中出现了两个表示因果关系的词。该句果在前、因在后,因而翻译宜采用倒译策略。这样既符合先果后因的汉语习惯,又层层递进、逻辑清晰。

Reinforced Learning

I. Answer the following questions for a comprehension of the text.

1. What are the three major categories of the primary products of petroleum refining, and give some examples to each category.

2. What's the reason for the high value of shipments per employee in refining industry?

3. What were the reasons for the considerable economic pressure that the petroleum refining industry in the U. S. has felt?

4. Why did the increases in refinery outputs happen when refineries closed?

5. Please explain that demand for refined petroleum products is expected to increase slowly with the growth of the U. S. economy.

II. Multiple choice: choose the correct one from the alternative answers to give the exact meaning of the word underlined.

1. Petroleum refining is one of the leading manufacturing industries in the United States in terms of its share of the total value of shipments of the

U. S. economy.

    A. exports   B. ships   C. carriage   D. imports

    2. In relation to its economic importance, however, the industry is comprised of relatively few companies and facilities.

    A. institutions   B. sectors   C. companies   D. infrastructure

    3. The primary products of the industry fall into three major categories: fuels (motor gasoline, diesel and distillate fuel oil, liquefied petroleum gas, jet fuel, residual fuel oil, kerosene, and coke); finished nonfuel products (solvents, lubricating oils, greases, petroleum wax, petroleum jelly, asphalt, and coke); and chemical industry feedstocks (naphtha, ethane, propane, butane, ethylene, propylene, butylenes, butadiene, benzene, toluene, and xylene).

    A. main   B. original   C. elementary   D. element

    4. Small refineries with capacities below 50,000 barrels per day, however, do play a significant role in the industry, making up about half of all facilities, but only 14 percent of the total crude distillation capacity.

    A. amounts   B. ability   C. abilities   D. facilities

    5. However, the industry also indirectly employs a significant number of outside contractors for many refinery operations, both routine and non – routine.

    A. regular   B. habitual   C. average   D. ordinary

    6. The value of product shipments sold by refining establishments was estimated to be $136 billion in 1992.

    A. groups   B. organizations C. businesses   D. institutions

    7. The imported share of crude oil is expected to increase as U. S. demand for petroleum products increases and the domestic production of crude oil declines.

    A. decreases   B. shortens   C. downwards   D. worsens

    8. The market determined that there was surplus capacity and the margins dropped to encourage the closure of the least efficient capacity. Reflecting these pressures, numerous facilities have closed in recent years.

    A. profits   B. edges   C. degrees   D. yields

    9. Between 1982 and 1994, the number of U. S. refineries as determined by the Department of Energy dropped from 301 to 176.

    A. found   B. surveyed   C. caused   D. decided

    10. For example, industry estimates of the capital costs to comply with the

1990 Clean Air Act Amendments, which mandates specific product compositions are about 35 to 40 billion.

    A. enterprise   B. firm   C. trade   D. insider

Ⅲ. Multiple choice: read the four suggested translations and choose the best answer.

    1. The primary products of the industry <u>fall into</u> three major categories: fuels (motor gasoline, diesel and distillate fuel oil, liquefied petroleum gas, jet fuel, residual fuel oil, kerosene, and coke); finished nonfuel products (solvents, lubricating oils, greases, petroleum wax, petroleum jelly, asphalt, and coke); and chemical industry feedstocks (naphtha, ethane, propane, butane, ethylene, propylene, butylenes, butadiene, benzene, toluene, and xylene).

    A. 分为   B. 落入   C. 陷于   D. 注入

    2. A relatively small number of people are employed by the petroleum refining industry <u>in relation to</u> its economic importance.

    A. 与……相比 B. 涉及   C. 鉴于   D. 就……而言

    3. The petroleum refining industry in the U. S. has felt considerable economic pressures in the past decade arising from a number of factors including: increased costs of labor; <u>compliance with</u> new safety and environmental regulations; and the elimination of government subsidies through the Crude Oil Entitlements.

    A. 顺从   B. 屈从   C. 遵守   D. 依从

    4. Some larger facilities, however, have also closed <u>in response</u> to economic pressures.

    A. 回应   B. 顺应   C. 响应   D. 应对

    5. Increases in refinery outputs are attributable to higher utilization rates of refinery capacity, and to incremental additions to the refining capacity at existing facilities as <u>opposed</u> to construction of new refineries.

    A. 与……对照 B. 与……相比 C. 与……对立   D. 与……相反

Ⅳ. Put the following sentences into Chinese.

    1. The value of product shipments sold by refining establishments was estimated to be $ 136 billion in 1992.

    2. Most U. S. crude oil distillation capacity is owned by large, integrated companies with multiple high capacity refining facilities.

3. In 1994, imports accounted for more than 50 percent of the crude oil used in the U. S. and about 10 percent of finished petroleum products.

4. Imported finished petroleum products serve specific market niches arising from logistical considerations, regional shortages, and long – term trade relations between suppliers and refiners.

5. A rationalization period began after crude oil pricing and entitlements were decontrolled in early 1981.

V. Put the following paragraphs into Chinese.

1. Petroleum refining is one of the leading manufacturing industries in the United States in terms of its share of the total value of shipments of the U. S. economy. In relation to its economic importance, however, the industry is comprised of relatively few companies and facilities. The number of refineries operating in the U. S. can vary significantly depending on the information source.

2. Demand for refined petroleum products is expected to increase slowly with the growth of the U. S. economy. The rate of increase will average about 1. 5 percent per year, which is slower than the expected growth of the economy. This slower rate of increase of demand will be due to increasing prices of petroleum products as a result of the development of substitutes for petroleum products, and rising costs of compliance with environmental and safety requirements.

1. 4   The Development of Refining Industry and Petroleum Products

🔲 Guidance to Reading

Petroleum refining and product development depends on the market demand. The new way of life, the increased needs for illuminants, for fuel to drive the factories, for gasoline to power the automobiles, all contributed to the increased use of petroleum and led to innovations and developments in the refining industry, thereby giving birth to the integrated petroleum refinery. However, unconventional petroleum refining remains a big challenge.

Chapter 1　American Refinery Industry Development

Text

1. The Development of Refining Industry

Petroleum in the **unrefined** or crude form, like many industrial feedstocks has little or no direct use and its value as an industrial commodity is only realized after the production of salable products. Even then, the market demand dictates the type of products that are needed. Therefore, the value of petroleum is directly related to the yield of products and is subject to the call of the market.

Petroleum refining, also called petroleum processing, is the recovery and/or generation of usable or salable fractions and products from crude oil, either by distillation or by chemical reaction of the crude oil **constituents** under the effects of heat and pressure. Synthetic crude oil, produced from **tar sand (oil sand) bitumen**, is also used as feedstocks in some refineries. As the basic elements of crude oil, hydrogen and carbon form the main input into a refinery, combining into thousands of individual constituents and the economic recovery of these constituents varies with the individual petroleum according to its particular individual qualities, and the processing facilities of a particular refinery.

In general, crude oil, once refined, yields three basic groupings of products that are produced when it is separated into a variety of different generic, but often **overlapping** fractions. The amounts of these fractions produced by distillation depend on the origin and properties of crude petroleum. The gas and gasoline cuts form the lower boiling products and are usually more valuable than the higher boiling fractions and provide gas (liquefied petroleum gas), naphtha, **aviation fuel**, **motor fuel**, and feedstocks, for the petrochemical industry. Naphtha, a precursor to gasoline and **solvents**, is extracted from both the light and middle range of distillate cuts and is also used as a feedstock for the petrochemical industry. **The middle distillates** refer to products from the middle boiling range of petroleum and include kerosene, diesel fuel, distillate fuel oil, and **light gas oil**; waxy distillate and lower boiling **lubricating oils** are sometimes included in the middle distillates. The remainder of the crude oil includes the higher boiling lubricating oils, **gas oil**, and **residuum** (the nonvolatile fraction of the crude oil). The residuum can also produce heavy lubricating oils and waxes, but is more often used for asphalt production. The complexity of petroleum is emphasized in so far as the actual proportions of light, medium, and heavy fractions vary significantly from one crude oil to another.

When petroleum occurs in a reservoir that allows the crude material to be recovered by pumping operations as a free – flowing dark – to – light colored liquid, it is often referred to as conventional petroleum. In some oil fields, the down hole pressure is sufficient for recovery without the need for pumping. **Heavy oil** differs from conventional petroleum in that its flow properties are reduced and it is much more difficult to recover from the subsurface reservoir. These materials have a much higher **viscosity** and lower **API** (**American Petroleum Institute**) **gravity** than conventional petroleum, and primary recovery of these petroleum types usually requires thermal stimulation of the reservoir.

Heavy oil generally has an API gravity of < 20 and usually, but not always, a sulfur content of > 2% by weight. **Extra heavy oil** occurs in the near – solid state and is virtually incapable of free flow under **ambient** conditions. Tar sand bitumen, often referred to as native asphalt, is a subclass of extra heavy oil and is frequently found as the organic filling in **pores** and **crevices** of **sandstones, limestones**, or **argillaceous sediments**, in which case the organic and **associated mineral matrix** is known as rock asphalt. A residuum, often shortened to resid, is the residue obtained from petroleum after nondestructive distillation has removed all the **volatile** materials.

The temperature of the distillation is usually < 345℃ because the rate of thermal **decomposition** of petroleum constituents is substantially > 350℃. Temperatures as high as 425℃ can be employed in **vacuum distillation**. When such temperatures are employed and thermal decomposition occurs, the residuum is usually referred to as **pitch**. Asphalt, prepared from petroleum, often resembles native asphalt. When asphalt is produced by distillation, the product is called residual asphalt. However, if the asphalt is prepared by solvent **extraction** of residua or by light hydrocarbon (propane) precipitation, or if it is blown or otherwise treated, the name should be modified accordingly to qualify the product, eg, propane asphalt. **Sour** and **sweet** are terms referring to a crude oil's approximate sulfur content. A crude oil that has a high sulfur content usually contains **hydrogen sulfide**, H_2S, and/or **mercaptans**, RSH; it is called sour.

2. The Development of Petroleum Products

The use of petroleum or **derived materials**, such as asphalt, and the heavier nonvolatile crude oils is an old art. In fact, petroleum utilization has been documented for >5000 years. The earliest documented uses occurred in Mesopotamia (ancient Iraq) when it was recognized that the nonvolatile derivatives (bitumen or natural asphalt and manufactured asphalt) could be used for **caulking** and as an **adhesive** for jewelry or as a mastic for construction purposes. Approximately 2000 years ago, Arabian scientists developed methods for the distillation of petroleum, which were introduced into Europe by way of the Arabian incursions into Spain. Interest in naphtha (nafta) began with the discovery that petroleum could be used as an illuminant and as a supplement to bituminous incendiaries, which were becoming increasingly common in warfare. Greek fire was a naphtha – bitumen (or naphtha – asphalt) mix; the naphtha provided the flame and the bitumen (or asphalt) provided the adhesive properties that prolonged the incendiary effect.

Modern refining began in 1859 with the discovery of petroleum in Pennsylvania. After completion of the first well, the surrounding areas were immediately leased and extensive drilling took place. The impetus to develop the petroleum refining industry came from several changes in life – styles. The increased needs for illuminants, for fuel to drive the factories of the industrial revolution, for gasoline to power the automobiles, as well as the demand for aviation fuel, all contributed to the increased use of petroleum. The product slate has also changed. The increased demand for gasoline and lubricants brought about an emphasis on refining crude oil. This, in turn, brought about changes in the way crude oil was refined and led to innovations and developments in the refining industry, thereby giving birth to the integrated petroleum refinery.

The constant demand for products, such as liquid fuels, is the main driving force behind the petroleum industry. In general, when the product is a fraction that has been produced from petroleum and includes a large number of individual hydrocarbons, the fraction is classified as a refined product. Examples of refined products are gasoline, diesel fuel, heating oils, lubricants, waxes, asphalt, and coke. In contrast, when the product is limited to, perhaps, one or two specific hydrocarbons of high purity, the fraction is classified as a petrochemical product.

Certain specific hydrocarbons, such as propane, butane, pentane, and their mixtures, exist in the gaseous state under atmospheric ambient conditions, but can be converted to the liquid state under conditions of moderate pressure at ambient temperature. This is termed **LPG** (liquefied petroleum gas), a refinery product.

Words and Expressions

unrefined	adj. 未经炼制的
constituent	n. 组分
bitumen	n. 地沥青
overlapping	adj. 重叠的
residuum	n. 渣油
viscosity	n. 黏度
ambient	adj. 周围的
pore	n. 孔隙
crevice	n. 裂缝
sandstone	n. 砂岩
limestone	n. 石灰岩
volatile	adj. 挥发性的
decomposition	n. 分解
pitch	n. 沥青
extraction	n. 萃取
precipitation	n. 沉淀
sour	adj. 高硫的
sweet	adj. 低硫的
mercaptan	n. 硫醇
caulk	v. 使密封防水
adhesive	n. 黏合剂

Technical Terms

petroleum refining	石油炼制
tar sand	焦油砂
oil sand	油砂
the middle distillate	中间馏分油
light gas oil	轻瓦斯油
lubricating oil	润滑油

gas oil	瓦斯油
heavy oil	重油
API gravity	美国石油学会比重指数,API 度
extra heavy oil	超重油
argillaceous sediments	泥质沉积物
associated mineral matrix	伴生矿物基质
vacuum distillation	减压蒸馏
hydrogen sulfide	硫化氢
derived material	衍生物
LPG	液化石油气

Language Focus

1. The Development of Refining Industry and Petroleum Products.

本章节题目中的"development"同时与"industry"和"products"搭配,采取名词并列形式可使题目简洁明了。翻译过程中,译者对其进行了增译处理,译为"炼制业的发展和石油产品开发"。

2. Petroleum in the unrefined or crude form, like many industrial feedstocks, has little or no direct use and its value as an industrial commodity is only realized after the production of salable products.

[参考译文为:石油(也称为原油),在未炼制或天然状态下,就像许多工业原料一样几乎很少或者不能直接使用,其工业商品的价值只能在转化为销售产品后才能实现。]

本句中"in...form"表示"以······的形式或状态","Petroleum in the unrefined or crude form..."意为:石油在未炼制或天然状态下。"like many industrial feedstocks,"为插入语,意为:就像许多工业原料一样。

3. Even then, the market demand dictates the type of products that are needed.

"even then"和"even so/now"都是"虽然如此"的意思。"dictate"一词本意为"规定,命令,支配",此处为拟人化用法,将市场看作可以发号施令的人,使文章更加生动形象。该句的翻译可以采用正译法,译为"虽然如此,市场需求仍决定着所需产品的种类";亦可采用反译法,译为"即便如此,产品类型仍然取决于市场的需求"。

4. Naphtha, a precursor to gasoline and solvents, is extracted from both the light and middle range of distillate cuts and is also used as a feedstock for the

petrochemical industry.

（参考译文为：石脑油，即汽油和溶剂的前身，是从轻质和中质馏分油中提取的，也是石化工业的原料。）

此句中有两个"and"，第一个"and"并列"light"和"middle"修饰"range"，第二个"and"并列"is extracted..."和"is also used..."。

5. The use of petroleum or derived materials, such as asphalt, and the heavier nonvolatile crude oils is an old art.

（参考译文为：石油及其衍生物的使用是一种古老的工艺，如沥青和较重的非挥发性原油。）

本句的核心结构是"the use... is an old art"。句中"art"意为"工艺"，在科技英语中常用。

Reinforced Learning

I. Answer the following questions for a comprehension of the text.

1. What are the differences between heavy oil and conventional petroleum?

2. What's the definition of tar sand bitumen?

3. What are the different names for asphalt and the naming reasons?

4. How did the life – styles foster the development of petroleum refining industry?

5. What are the differences between refined product and petrochemical product?

II. Multiple choice: choose the correct one from the alternative answers to give the exact meaning of the word underlined.

1. Even then, the market demand <u>dictates</u> the type of products that are needed.

A. decrees   B. controls   C. orders   D. records

2. As the basic elements of crude oil, hydrogen and carbon form the main <u>input</u> into a refinery, combining into thousands of individual constituents and the economic recovery of these constituents varies with the individual petroleum according to its particular individual qualities, and the processing facilities of a particular refinery.

A. ingredients   B. investment   C. money   D. products

3. Heavy oil differs from conventional petroleum in that its flow <u>properties</u>

are reduced and it is much more difficult to recover from the subsurface reservoir.

A. qualities　　B. possessions　　C. speeds　　D. effects

4. Extra heavy oil occurs in the near – solid state and is virtually incapable of free flow under ambient conditions.

A. almost　　B. really　　C. truly　　D. imaginarily

5. The temperature of the distillation is usually <345℃ because the rate of thermal decomposition of petroleum constituents is substantially >350℃.

A. considerably　B. mainly　　C. truly　　D. nearly

6. In fact, petroleum utilization has been documented for >5000 years.

A. supported　　B. proven　　C. recorded　　D. used

7. Interest in naphtha (nafta) began with the discovery that petroleum could be used as an illuminant and as a supplement to bituminous incendiaries, which were becoming increasingly common in warfare.

A. replacement　B. complement　　C. assistance　　D. necessity

8. After completion of the first well, the surrounding areas were immediately leased and extensive drilling took place.

A. massive　　B. comprehensive C. abundantly　　D. complicated

9. The impetus to develop the petroleum refining industry came from several changes in life – styles.

A. stimulus　　B. factors　　C. motivations　D. intention

10. The increased needs for illuminants, for fuel to drive the factories of the industrial revolution, for gasoline to power the automobiles, as well as the demand for aviation fuel, all contributed to the increased use of petroleum.

A. operate　　B. motivate　　C. push　　D. force

Ⅲ. Multiple choice：read the four suggested translations and choose the best answer.

1. Therefore, the value of petroleum is directly related to the yield of products and is subject to the call of the market.

A. ……的吸引　B. ……的呼唤　　C. ……的要求　D. ……的命令

2. The amounts of these fractions produced by distillation depend on the origin and properties of crude petroleum.

A. 依靠　　B. 确信　　C. 取决于　　D. 借助

3. The complexity of petroleum is emphasized in so far as the actual pro-

portions of light, medium, and heavy fractions vary significantly from one crude oil to another.

    A. 在……范围内   B. 到……程度

    C. 迄今为止   D. 如此……以至于

    4. Asphalt, prepared from petroleum, often resembles native asphalt.

    A. 提取   B. 调到   C. 准备   D. 筹备

    5. In general, when the product is a fraction that has been produced from petroleum and includes a large number of individual hydrocarbons, the fraction is classified as a refined product.

    A. 单个的   B. 个别的   C. 独特的   D. 个人的

Ⅳ. Put the following sentences into Chinese.

    1. Synthetic crude oil, produced from tar sand (oil sand) bitumen, is also used as feedstocks in some refineries.

    2. In general, crude oil, once refined, yields three basic groupings of products that are produced when it is separated into a variety of different generic, but often overlapping fractions.

    3. The complexity of petroleum is emphasized in so far as the actual proportions of light, medium, and heavy fractions vary significantly from one crude oil to another.

    4. In some oil fields, the down hole pressure is sufficient for recovery without the need for pumping.

    5. Greek fire was a naphtha - bitumen (or naphtha - asphalt) mix; the naphtha provided the flame and the bitumen (or asphalt) provided the adhesive properties that prolonged the incendiary effect.

V. Put the following paragraphs into Chinese.

    1. As the basic elements of crude oil, hydrogen and carbon form the main input into a refinery, combining into thousands of individual constituents and the economic recovery of these constituents varies with the individual petroleum according to its particular individual qualities, and the processing facilities of a particular refinery.

    2. Certain specific hydrocarbons, such as propane, butane, pentane, and their mixtures, exist in the gaseous state under atmospheric ambient conditions, but can be converted to the liquid state under conditions of moderate pressure at ambient temperature. This is termed LPG (liquefied petroleum gas), a refinery product.

Chapter 2　Petroleum Refining Processes

2. 1　Introduction to Petroleum

⊡ Guidance to Reading

How does petroleum originate? What is it made up of? The evidence supporting a biological source for the material that generates petroleum is extensive. The generation of petroleum is nonbiological, induced by temperature, and influenced by available time. For petroleum accumulation, migration has a critical role in linking the organic – rich source rocks to the reservoir. The composition of petroleum changes and evolves in the reservoir in response to changing conditions.

⊡ Text

Origin of Petroleum

Petroleum is a naturally occurring complex mixture made up predominantly of **carbon and hydrogen compounds**, but also frequently containing significant amounts of nitrogen, sulfur, and oxygen together with smaller amounts of **nickel, vanadium**, and other elements. Because of their fluid nature petroleum **phases** are mobile in the subsurface and may have **accumulated** far from the place where they formed, present as commercial accumulations in **permeable** and **porous** reservoir rocks. A biogenic origin for the **carbonaceous material** in petroleum is widely but not universally accepted. However, oils contain the so – called chemical fossils or biomarkers, compounds having characteristic molecular structures that can be related to living systems. The compounds include isoprenoids, porphyrins, steranes, hopanes, and many others. The relative abundances of members of homologous series are often similar to those in living systems. In addition, the lack of **thermodynamic equilibrium** among compounds, and the close association of petroleum with **sedimentary rocks** formed in an **aqueous** environment, suggests a low temperature (less than a few hundred degrees Celsius) origin. The elemental composition of petroleum (C, H, N, S, O), the isotopic composition of oils, and the presence of petroleum – like materials in more recent sediments are consistent with a low temperature origin. The evidence supporting a biological source for the material that generates petroleum is extensive.

Organisms produce a wide range of organic compounds including significant amounts of **biopolymers** like proteins, **carbohydrates**, and **lignins**, together with a wide variety of lower **molecular weight lipids**. After the death of the organism, all or part of this organic material may accumulate in aquatic environments where the various compounds have very different stabilities. Some are **metabolized** in the water column by other organisms (including bacteria) and only the **biochemically** resistant material is incorporated into sediments. Survival of organic material depends on many factors but particularly the **oxidizing** or reducing nature of the system. **Preservation** is strongly favored in **anoxic sediments**. However, the formation of a petroleum accumulation requires more than just a concentration of the relatively low molecular weight hydrocarbons that are present in more recent sediments. Although $C_2 \sim C_{10}$ hydrocarbons are present in extremely low [parts per billion (ppb) level] concentrations in organisms and sediments, these can account for up to 50% or more of the volume of some crude oils.

The generation of petroleum is **nonbiological**, induced by temperature, and influenced by available time. It follows essentially first – order kinetics where an increase of 10℃ roughly doubles the reaction rate at low temperatures. The low molecular weight compounds generated from the **kerogen** show none of the biological characteristics typical of compounds in more recent sediments. Therefore, at increasing depth, and hence increasing temperature, the bitumen fraction loses features such as **odd** – **even** predominance in **long – chain normal alkanes, optical activity**, and the predominance of four – and five – ringed **cycloalkanes**. These trends have been well – documented in many areas. The petroleum generation process can be treated **quantitatively** using models based on first – order kinetics.

Most petroleum is found in reservoir rocks that have high **permeabilities** and **porosities**, where these properties have been developed by natural processes of sorting and winnowing that remove fine – grained particles, including organic materials. Reservoir rocks generally have insufficient organic matter to generate commercially significant quantities of petroleum. It is believed that petroleum generation occurs in organic – rich **source rocks**, and that part of the bitumen then migrates to accumulate in **reservoir rocks**. This is the source rock concept. Clearly, migration has a critical role in linking the organic – rich source rocks to the reservoir. **Solubilities** of hydrocarbons in subsurface waters are generally too low to be significant in the petroleum migration process, and most recent studies have stressed the **expulsion** of a separate crude oil phase out of source rocks. It appears that source rocks develop high internal petroleum saturation caused by petroleum generation and the **displacement** of water during **compaction**. Upon continuing compaction and kerogen conversion, oil droplets are forced out of source rocks into adjacent permeable **carrier beds**. High pressures approaching the rock load (lithostatic) can develop, and induce near – vertical fractures that are important in providing migration pathways out of the source rock. In this case the pressure gradient that develops can overcome buoyancy, and oil may be expelled downward and out the bottom of source rocks, as well as out of the top. The alkanes are less strongly absorbed in the source rock and so are preferentially expelled. In contrast, the nitrogen – sulfur – oxygen compounds (NSOs) are most strongly adsorbed and thus depleted in the expelled oil. Migration efficiency varies widely and appears to be dependent on the organic matter content of the rock. High efficiencies are associated with high organic contents. The controlling factor is hydrocarbon (bitumen) saturation, and rocks having less than about 1.0% (wt) organic matter do not generate sufficient bitumen. As a consequence no oil migrates from these rocks and they are not effective source rocks.

Buoyancy is the main driving force through the carrier beds and oils continue to move upward until stopped in a **structural trap**, or where permeability decreases as in a stratigraphic trap. Migration distances can be in excess of 100 km. Oil may be remobilized after its initial accumulation in the reservoir. Although in the simplest case this may involve only a simple relocation, it can lead to significant compositional changes if both gas and oil are involved. When an **anticlinal reservoir** is full to the spill point and has a **gas cap** over oil, any spilling off the bottom is oil, and the next shallower trap thus accumulates oil with no gas cap. This process of differential entrapment leads eventually to oil in the up dip (shallower) reservoirs and gas in the deeper ones. Geological examples are given.

The composition of petroleum changes and evolves in the reservoir in response to changing conditions. Thermal maturation of crude oil is brought about by the increasing temperature that accompanies increasing depth of burial. Some large molecules are broken down into smaller fragments and the trend is for an increasing percentage of the lighter fractions as the oil progresses to lower densities in the sequence from oil, to lighter oil, to wet gas, and finally dry gas. The increasing hydrogen content implied by this sequence is provided by parallel reactions involving **cyclization** and **aromatization**. These residual molecules get steadily more **aromatic** and larger, and as the solvent properties of the oil change, these molecules are precipitated in the reservoir in a process called natural **deasphaltening**. The solid precipitate also takes with it much of nickel and vanadium, so that the producible oil is of better quality.

Words and Expressions

nickel	n. 镍
vanadium	n. 钒
phase	n. 相,位
accumulate	v. 聚集
permeable	adj. 可渗透的
porous	adj. 有孔的
aqueous	adj. 含水的
biopolymer	n. 生物聚合物
carbohydrate	n. 碳水化合物
lignin	n. 木质素
lipid	n. 脂类
metabolize	v 代谢作用
biochemically	adv. 生物化学地
oxidize	v. 氧化
preservation	n. 防腐
anoxic	adj. 厌氧的
sediment	n. 沉积物
nonbiological	adj. 非生物化的
kerogen	n. 干酪根,油母质

odd – even	n. 奇偶性
cycloalkane	环烷烃
quantitatively	adv. 定量地，量化地
permeability	n. 渗透率
porosity	n. 孔隙率
solubility	n. 溶解度
expulsion	n. 排斥
displacement	n. 驱替
compaction	n. 挤压
cyclization	n. 环化
aromatization	n. 芳构化
aromatic	n. 芳族
deasphalten	v. 脱沥青

Technical Terms

carbon and hydrogen compound	碳水化合物
carbonaceous material	含碳物质
thermodynamic equilibrium	动力平衡
sedimentary rock	沉积岩
molecular weight	相对分子质量
long – chain normal alkane	长链正构烷烃
optical activity	光学活性
source rock	生油岩
reservoir rock	储油岩
carrier bed	运载岩
structural trap	构造圈闭
anticlinal reservoir	背斜油藏
gas cap	气顶

Language Focus

1. Because of their fluid nature petroleum phases are mobile in the subsurface and may have accumulated far from the place where they formed, present as commercial accumulations in permeable and porous reservoir rocks.

（参考译文为：由于天然液态石油在地层下是流动的，可能在距离发源地很远的地方汇聚。鉴于此,石油的形成是一个复杂的过程,起始于通过自然运动,最终在可渗透多孔储集层中聚集,并以油气藏方式存在。)

连词"and"并列"are mobile"和"may have accumulated...","where"引导的定语从句修饰先行词"the place",说明石油形成的地方和聚集的地方往往不一致。

2. The relative abundances of members of homologous series are often similar to those in living systems.

(参考译文为:石油中相对丰富的同类物常与生态系统中的该类物质相似。)

句中"those"意为在生态系统中也存在的"组成生命体的成分"。

3. The evidence supporting a biological source for the material that generates petroleum is extensive.

该句的主体框架为"the evidence is extensive","extensive"意为"广泛的,大量的",意在表示:大量证据表明石油来自于生物。另外,句中"supporting"为现在分词作定语;"that"引导定语从句修饰先行词"material"。

4. The generation of petroleum is nonbiological, induced by temperature, and influenced by available time.

(参考译文为:石油的生成是非生物过程,由温度诱发,而且受有效时间影响。)

句中"and"为连接词,起递进作用,译为"而且"。

5. When an anticlinal reservoir is full to the spill point and has a gas cap over oil, any spilling off the bottom is oil, and the next shallower trap thus accumulates oil with no gas cap.

(参考译文为:当背斜油藏达到溢点并且原油以上有气顶时,任何从底部的泄漏都是原油泄漏,因此下一个浅层圈闭积聚原油的时候就不会伴随有气顶出现。)

该句中动词全都采用一般现在时,说明科技文章是对事实的描写,客观而真实。"thus"一词表明该句中存在因果关系,先因后果更符合科技文体的叙述特点。

🔲 Reinforced Learning

I. Answer the following questions for a comprehension of the text.

1. What does petroleum made up of?

2. How did the organism get into sediments after death?

3. In what way can we say that the generation of petroleum follows essentially first – order kinetics?

4. What's the role of migration?

5. How does buoyancy affect migration?

6. How does the thermal maturation of crude oil take place?

II. Multiple choice : choose the correct one from the alternative answers to give the exact meaning of the word underlined.

1. In addition, the lack of thermodynamic equilibrium among compounds, and the close association of petroleum with sedimentary rocks formed in an aqueous environment, <u>suggests</u> a low temperature (less than a few hundred degrees Celsius) origin.

    A. shows   B. proposes   C. arouses   D. implies

2. Organisms produce a wide range of organic compounds including <u>significant</u> amounts of biopolymers like proteins, carbohydrates, and lignins, together with a wide variety of lower molecular weight lipids.

    A. expressive   B. meaningful   C. considerable   D. important

3. Preservation is strongly <u>favored</u> in anoxic sediments.

    A. supported   B. helped   C. approved   D. accepted

4. Although $C_2 \sim C_{10}$ hydrocarbons are <u>present</u> in extremely low [parts per billion (ppb) level] concentrations in organisms and sediments, these can account for up to 50% or more of the volume of some crude oils.

    A. existed   B. described   C. appeared   D. showed

5. It follows <u>essentially</u> first – order kinetics where an increase of 10℃ roughly doubles the reaction rate at low temperatures.

    A. basically   B. necessarily   C. indispensably   D. importantly

6. The petroleum generation process can be <u>treated</u> quantitatively using models based on first – order kinetics.

    A. handled   B. developed   C. considered   D. described

7. Clearly, migration has a <u>critical</u> role in linking the organic – rich source rocks to the reservoir.

    A. key   B. basic   C. valuable   D. fair

8. Although in the simplest case this may involve only a simple relocation, it can lead to significant compositional changes if both gas and oil are <u>involved</u>.

    A. included   B. existed   C. affected   D. absorbed

9. The composition of petroleum changes and <u>evolves</u> in the reservoir in response to changing conditions.

    A. changes   B. develops   C. grows   D. increases

10. The increasing hydrogen content <u>implied</u> by this sequence is provided by parallel reactions involving cyclization and aromatization.

    A. meant   B. read   C. expressed   D. hinted

III. Multiple choice：read the four suggested translations and choose the best answer.

1. However, oils contain the so – called chemical fossils or biomarkers, compound having characteristic molecular structures that can <u>be related to</u> living systems.

    A. 与……相关   B. 涉及   C. 针对   D. 适用于

2. Some are metabolized in the water column by other organisms (including bacteria) and only the biochemically resistant material is <u>incorporated into</u> sediments.

    A. 被吸收   B. 被包含   C. 被合并   D. 被融入

3. Migration distances can be <u>in excess of</u> 100 km. Oil may be remobilized after its initial accumulation in the reservoir.

    A. 超过   B. 超额   C. 过度   D. 过量

4. When an anticlinal reservoir <u>is full to</u> the spill point and has a gas cap over oil, any spilling off the bottom is oil, and the next shallower trap thus accumulates oil with no gas cap.

    A. 达到……的   B. 充足的   C. 充满的   D. 完全的

5. The composition of petroleum changes and evolves in the reservoir <u>in response</u> to changing conditions.

    A. 应对   B. 依据   C. 随着   D. 适应

IV. Put the following sentences into Chinese.

1. A biogenic origin for the carbonaceous material in petroleum is widely but not universally accepted.

2. The relative abundances of members of homologous series are often similar to those in living systems.

3. The alkanes are less strongly absorbed in the source rock and so are preferentially expelled.

4. Although in the simplest case this may involve only a simple relocation, it can lead to significant compositional changes if both gas and oil are involved.

5. Thermal maturation of crude oil is brought about by the increasing temperature that accompanies increasing depth of burial.

V. Put the following paragraphs into Chinese.

1. Petroleum is a naturally occurring complex mixture made up predominantly of carbon and hydrogen compounds, but also frequently containing significant amounts of nitrogen, sulfur, and oxygen together with smaller amounts of nickel, vanadium, and other elements. Because of their fluid nature petroleum phases are mobile in the subsurface and may have accumulated far from the place where they formed, present as commercial accumulations in permeable and porous reservoir rocks.

2. Organisms produce a wide range of organic compounds including significant amounts of biopolymers like proteins, carbohydrates, and lignins, together with a wide variety of lower molecular weight lipids. After the death of the organism, all or part of this organic material may accumulate in aquatic environments where the various compounds have very different stabilities. Some are metabolized in the water column by other organisms (including bacteria) and only the biochemically resistant material is incorporated into sediments.

2.2 Petroleum Composition

Guidance to Reading

The molecular composition of the liquid portion of petroleum contributes to the crude oil properties and behavior. Physical properties such as boiling point, density, odor, and viscosity have been used to classify oils. Knowledge of the composition of petroleum allows the refiner to optimize conversion of raw petroleum into high value products, and environmentalists to consider the biological impact of environmental exposure to avoid wasteful production and pollution to the environment.

Text

Petroleum can include three phases: gaseous (natural gas), liquid (crude oil), and solid or semisolid (bitumens, asphalt, tars, and pitches). The molecular composition of the liquid portion of petroleum contributing to the crude oil properties and behavior is discussed in this section. Crude oils vary dramatically in color, **odor**, and flow properties. These properties often reflect the origin of the crude. Historically, physical properties such as boiling point, density (gravity), odor, and **viscosity** have been used to classify oils. Crude oils may be called light or heavy in reference to relative density. Light crude oils are rich in low boiling and **paraffinic** hydrocarbons; heavy crude oils contain greater amounts of high boiling and asphalt – like molecules. The heavy oils tend to be more viscous, higher boiling, more aromatic, and contain larger amounts of **heteroatoms**. Likewise, odor is used to distinguish between sweet or low sulfur, and sour or high sulfur, crude oils.

Knowledge of the composition of petroleum allows the refiner to optimize conversion of raw petroleum into high value products. Originally, petroleum was distilled and sold as fractions, primarily for use in illumination and lubrication. Crude oil is sold in the form of gasoline, solvents, diesel and jet fuel, heating oil, lubricant oils, and asphalts, or it is converted to petrochemical feedstocks such as **ethylene, propylene**, the **butenes**, butadiene, and isoprene. Modern refining uses a sophisticated combination of heat, **catalyst**, and hydrogen. Conversion processes include **coking, hydrocracking**, and **catalytic cracking** to break large molecules into smaller fractions; hydrotreating to reduce heteroatoms and aromatics, thereby creating environmentally acceptable products; and **isomerization** and reforming to rearrange molecules to those having high value, eg, gasolines of high octane number.

Knowledge of the molecular composition of petroleum also allows environmentalists to consider the biological impact of environmental exposure. Increasingly, petroleum is being produced in and transported from remote areas of the world to refineries located closer to markets. Although only a minuscule fraction of that oil is released into the environment, the sheer volume involved has the potential for environmental damage. Molecular composition can not only identify the sources of contamination but also aids in understanding the fate and effects of the potentially hazardous components.

Crude oils contain an extremely wide range of organic functionality and molecular size. The variety is so great that a complete compound-by-compound description for even a single crude oil is not likely. The molecular composition of petroleum can, however, be described in terms of three classes of compounds: saturates, **aromatics**, and compounds bearing the heteroatoms sulfur, oxygen, or nitrogen. Within each of these classes there are several families of related compounds.

Molecular Classes. The molecules in crude oil include several basic structural types. Because they may contain from 1 to 100 carbon atoms and may occur in combination, the statistical potential for isomeric structures is staggering. For example, whereas there are just 75 possible paraffinic structures for C_{10}, there are $>10^5$ **isomers** for C_{20}. A few structures tend to dominate the distributions of each isomer group, however.

The inclusion of **naphthene** and other aromatic rings introduces two additional dimensions, increasing the number of hydrocarbon isomers even further. A three – dimensional **array** in which the molecules could be described in terms of the number of aromatic rings, the number of naphthenic rings, and the number of carbons in **alkyl side chains** has been proposed. Conceptually, this amounts to describing a three-dimensional molecular mountain for hydrocarbons. There is also the potential of constructing similar mountains for **heterocyclics**. The two-dimensional image in the naphthenic and aromatic dimensions has been projected using sidebars to indicate the variation in alkyl substituents among crude oils.

Molecular characterization of whole oil is beyond the capability of most analytical techniques. Distillation, however, can separate petroleum into molecular weight fractions that simplify the task. Pioneering work with this approach, sponsored by the American Petroleum Institute (API) starting in 1925, has led to the identification of hundreds of individual compounds in distillation fractions of a single crude oil. Developments in **chromatography** allowed oils to be **fractionated** by **polarity** as a second dimension. Under API sponsorship, the U. S. Bureau of Mines extended **separations** and measurement techniques to heavier fractions. At the same time, individual compounds have been isolated and quantified from increasingly higher boiling fractions. Techniques have been developed that use combinations of classical **open-column adsorption chroma-**

tography, **gel permeation chromatography**, and **ion-exchange** separations to i-solate fractions in which compounds could be identified by **mass spectrometry**.

Crude distillations yield different quantities in each fraction. About the same amounts are distilled into the middle distillate and **vacuum gas oil** from conventional crude oils. More naphtha is distilled from light crude oils and more **vacuum residuum** is obtained from heavy crude oils. The typical distribution of classes of petroleum compounds shows a significant shift with boiling point. Whereas the lower boiling fractions are dominated by **nonpolar** saturated hydrocarbons that exist in limited isomeric forms, the higher boiling fractions increasingly contain a larger variety of classes, that have, in turn, an increasing number of possible isomers. As the boiling point increases, aromatic ring structures build in, first as **naked rings**, then more and more as rings having attached side – chain and naphthene ring carbons. Polar compounds, typically those having O and N functionality, that appear as trace impurities in the lower boiling fractions gradually become significant components in the higher boiling fractions. This is confirmed by the distribution of S and N in petroleum against boiling point. The S, not including H_2S and the light sulfur compounds such as mercaptans and sulfides, present in petroleum gases, is more widely distributed than the nitrogen that concentrates in the highest boiling fraction. Not shown is the subtle decrease that occurs in H/C ratio with increasing boiling point reflecting the increasing number of aromatic ring types at higher boiling point. The metals, nitrogen, and oxygen are predominantly found in the higher boiling fractions rich in polars.

Analytical Approaches. Different analytical techniques have been applied to each fraction to determine its molecular composition. As the molecular weight increases, complexity increasingly shifts the level of analytical detail from quantification of most individual species in the naphtha to average molecular descriptions in the vacuum residuum. For the naphtha, classical techniques allow the isolation and identification of individual compounds by physical properties. Gas chromatographic (gc) resolution allows almost every compound having less than eight carbon atoms to be measured separately. The combination of gc with mass spectrometry (gc/ms) can be used for quantitation purposes when compounds are not well – resolved by gc.

Words and Expressions

odor	n. 气味
viscosity	n. 黏度
paraffinic	adj. 含石蜡的
heteroatom	n. 杂环原子
ethylene	n. 乙烯
propylene	n. 丙烯
butene	n. 丁烯
catalyst	n. 催化剂
coke	v. 焦化
hydrocrack	v. 加氢裂化
isomerization	n. 异构化
aromatic	n. 芳香烃
isomer	n. 异构体
naphthene	n. 环烷烃
array	n. 排列
alkyl	n. 烷基
heterocyclic	n. 杂环族化合物
chromatography	n. 色谱
fractionate	v. 分馏
polarity	n. 极性
separation	n. 分离
ion – exchange	n. 离子交换
nonpolar	adj. 非极性的

Technical Terms

catalytic crack	催化裂化
side chain	侧链
open – column adsorption chromatography	空心柱吸收色谱图
gel permeation chromatography	胶化渗透色谱图
mass spectrometry	质谱
vacuum gas oil	减压瓦斯油

vacuum residuum   减压渣油

naked ring   裸环

⊡ Language Focus

1. Likewise, odor is used to distinguish between sweet or low sulfur, and sour or high sulfur, crude oils.

"likewise"为副词,意为"同样地",修饰整个句子。比起句子,副词的使用可大大缩减篇幅,使文章言简意赅又能与上文很好地衔接,在科技英语中常见。

2. The S, not including H_2S and the light sulfur compounds such as mercaptans and sulfides, present in petroleum gases, is more widely distributed than the nitrogen that concentrates in the highest boiling fraction.

(参考译文为:石油气中的硫的分布不包括硫化氢和轻硫化合物,如硫醇类和硫化物,比高沸点馏分中氮的分布更广泛。)

此句以"more...than..."为标志,是比较状语从句中的差比句,经常表示两个不同主语的同一性质特征不同。本句的主干为"The S is more widely distributed than..."。

3. Not shown is the subtle decrease that occurs in H/C ratio with increasing boiling point reflecting the increasing number of aromatic ring types at higher boiling point.

(参考译文为:氢碳比没有随沸点升高而产生细微减小,表明在高沸点时芳香环种类会增多。)

此句为修辞倒装句(rhetorical inversion),强调氢碳比随沸点升高而变小的现象没有出现。

4. The combination of gc with mass spectrometry (gc/ms) can be used for quantitation purposes when compounds are not well – resolved by gc.

(参考译文为:当气相色谱法不能很好地进行定量分析时,可以将气相色谱法和质谱分析法组合使用。)

句中"quantitation"表示"定量分析"。

⊡ Reinforced Learning

I. Answer the following questions for a comprehension of the text.

1. What are the differences between heavy and light crude oil?

2. Which classes of compounds can be described in the molecular composition of petroleum?

3. Explain the three – dimensional array for molecules description.

4. What is the chromatography technology based on?

5. What are the differences of fractions from light and heavy crude oil?

II. Multiple choice: choose the correct one from the alternative answers to give the exact meaning of the word underlined.

1. Light crude oils <u>are rich in</u> low boiling and paraffinic hydrocarbons; heavy crude oils contain greater amounts of high boiling and asphalt – like molecules.

    A. have   B. contain   C. yield   D. are fertile of

2. Conversion processes include coking, hydrocracking, and catalytic cracking to break large molecules into smaller fractions; hydrotreating to reduce heteroatoms and aromatics, thereby creating environmentally <u>acceptable</u> products; and isomerization and reforming to rearrange molecules to those having high value, eg, gasolines of high octane number.

    A. agreeable   B. receivable   C. tolerate   D. satisfactory

3. Although only a minuscule fraction of that oil is released into the environment, the <u>sheer</u> volume involved has the potential for environmental damage.

    A. pure   B. complete   C. little   D. entire

4. Within each of these classes there are several <u>families</u> of related compounds.

    A. groups   B. members   C. ranks   D. categories

5. Because they may contain from 1 to 100 carbon atoms and may occur in combination, the statistical <u>potential</u> for isomeric structures is staggering.

    A. numbers   B. capabilities   C. qualities   D. properties

6. A three – dimensional array in which the molecules could be described in terms of the number of aromatic rings, the number of naphthenic rings, and the number of carbons in alkyl side chains has been <u>proposed</u>.

    A. suggested   B. presented   C. infirmed   D. entitled

7. Pioneering work with this approach, sponsored by the American Petroleum Institute (API) starting in 1925, has led to the <u>identification</u> of hundreds of individual compounds in distillation fractions of a single crude oil.

    A. improvement B. truth   C. result   D. discovery

8. The typical distribution of classes of petroleum compounds shows a significant shift with boiling point.

    A. exchange   B. change   C. transfer   D. develop

9. The metals, nitrogen, and oxygen are predominantly found in the higher boiling fractions rich in polars.

    A. mostly   B. mainly   C. basically   D. fundamentally

10. Different analytical techniques have been applied to each fraction to determine its molecular composition.

    A. decide   B. research   C. discover   D. analyze

III. Multiple choice: read the four suggested translations and choose the best answer.

1. Crude oils may be called light or heavy in reference to relative density.

    A. 参考   B. 依据   C. 提到   D. 借助

2. A three – dimensional array in which the molecules could be described in terms of the number of aromatic rings, the number of naphthenic rings, and the number of carbons in alkyl side chains has been proposed.

    A. 在……方面   B. 在……范围内
    C. 以……为依据   D. 在……期限内

3. Conceptually, this amounts to describing a three – dimensional molecular mountain for hydrocarbons.

    A. 共计为   B. 等于   C. 意味着   D. 相当于

4. Molecular characterization of whole oil is beyond the capability of most analytical techniques.

    A. 超出   B. 达不到   C. 超过   D. 不包括

5. As the boiling point increases, aromatic ring structures build in, first as naked rings, then more and more as rings having attached side – chain and naphthene ring carbons.

    A. 形成   B. 产生   C. 建成   D. 建立

IV. Put the following sentences into Chinese.

1. The variety is so great that a complete compound – by – compound description for even a single crude oil is not likely.

2. The inclusion of naphthene and other aromatic rings introduces two additional dimensions, increasing the number of hydrocarbon isomers even further.

3. At the same time, individual compounds have been isolated and quantified from increasingly higher boiling fractions.

4. About the same amounts are distilled into the middle distillate and vacuum gas oil from conventional crude oils.

5. The typical distribution of classes of petroleum compounds shows a significant shift with boiling point.

V. Put the following paragraphs into Chinese.

1. Knowledge of the molecular composition of petroleum also allows environmentalists to consider the biological impact of environmental exposure. Increasingly, petroleum is being produced in and transported from remote areas of the world to refineries located closer to markets. Although only a minuscule fraction of that oil is released into the environment, the sheer volume involved has the potential for environmental damage. Molecular composition can not only identify the sources of contamination but also aids in understanding the fate and effects of the potentially hazardous components.

2. The molecules in crude oil include several basic structural types. Because they may contain from 1 to 100 carbon atoms and may occur in combination, the statistical potential for isomeric structures is staggering. For example, whereas there are just 75 possible paraffinic structures for C_{10}, there are $> 10^5$ isomers for C_{20}. A few structures tend to dominate the distributions of each isomer group, however.

2.3 Petroleum Refining

🔲 Guidance to Reading

The composition of crude oil can vary significantly depending on its source. Petroleum refineries are a complex system of multiple operations and the operations used at a given refinery depend upon the properties of the crude oil to be refined and the desired products. Generally, Petroleum Refining includes two phases and a number of supporting operations.

🔲 Text

Crude oil is a mixture of many different hydrocarbons and small amounts of impurities. The composition of crude oil can vary significantly depending on its source. Petroleum refineries are a complex system of multiple operations and the operations used at a given refinery depend upon the properties of the crude oil to be refined and the desired products. For these reasons, no two refineries are alike. Portions of the outputs from some processes are refed back into the same process, fed to new processes, fed back to a previous process, or blended with other outputs to form finished products. The major unit operations typically involved at petroleum refineries are described briefly below. In addition to those listed below, there are also many special purpose processes that cannot be described here and which may play an important role in a facility's efforts to comply with pollutant discharge and product specification requirement.

Refining crude oil into useful petroleum products can be separated into two phases and a number of supporting operations. The first phase is **desalting** of crude oil and the subsequent distillation into its various components or "fractions". The second phase is made up of three different types of "downstream" processes: combining, breaking, and **reshaping**. Downstream processes convert some of the distillation fractions into petroleum products (residual fuel oil, gasoline, kerosene, etc.) through any combination of different cracking, coking, reforming, and **alkylation** processes. Supporting operations may include wastewater treatment, sulfur recovery, additive production, **heat exchanger** cleaning, blowdown systems, blending of products, and storage of products.

Crude Oil Distillation and Desalting

One of the most important operations in a refinery is the initial distillation of the crude oil into its various boiling point fractions. Distillation involves the heating, **vaporization**, fractionation, **condensation**, and cooling of feedstocks. This text discusses the **atmospheric and vacuum distillation** processes which when used in sequence result in lower costs and higher efficiencies. Also discussed here is the important first step of desalting the crude oil prior to distillation.

Desalting

Before separation into fractions, crude oil usually must first be treated to remove **corrosive** salts. The desalting process also removes some of the metals and **suspended solids** which cause **catalyst deactivation**. Desalting involves the mixing of heated crude oil with water (about 3 to 10 percent of the crude oil volume) so that the salts are dissolved in the water. The water must then be separated from the crude oil in a separating vessel by adding **demulsifier** chemicals to assist in breaking the **emulsion** and/or, more commonly, by applying a high **potential** electric field across the settling vessel to **coalesce** the polar salt water droplets. The desalting process creates an oily **desalter sludge** and a high temperature salt water waste stream which is typically added to other process wastewaters for treatment in the refinery wastewater treatment facilities. The water used in crude desalting is often untreated or partially treated water from other refining process water.

Atmospheric Distillation

The desalted crude oil is then heated in a heat exchanger and furnace to about 750 degrees ($^{\circ}F$) and fed to a vertical, **distillation column** at atmospheric pressure where most of the feed is vaporized and separated into its various fractions by **condensing** on 30 to 50 fractionation trays, each corresponding to a different condensation temperature. The **lighter fractions** condense and are collected towards the top of the column. **Heavier fractions**, which may not vaporize in the column, are further separated later by vacuum distillation. Within each atmospheric distillation tower, a number of **side streams** (at least four) of low – boiling point components are removed from the tower from different trays. These low – boiling point mixtures are in equilibrium with heavier components which must be removed. The side streams are each sent to a different small **stripping tower** containing 4 to 10 trays with steam injected under the bottom tray. The steam strips the light – end components from the heavier components and both the steam and light – ends are fed back to the atmospheric distillation tower above the corresponding side stream **draw tray**. Fractions obtained from atmospheric distillation include naphtha, gasoline, kerosene, light fuel oil, diesel oils, gas oil, lube distillate, and heavy bottoms. Most of these can be sold as finished products, or blended with products from downstream processes. Another product produced in atmospheric distillation, as well as many other refinery processes, is the light, noncondensible refinery fuel gas (mainly **methane** and **ethane**). Typically this gas also contains **hydrogen sulfide** and ammonia gases. The mixture of these gases is known as "**sour gas**" or "**acid gas**". The sour gas is sent to the refinery sour gas treatment system which separates the fuel gas so that it can be used as fuel in the refinery heating furnaces. Air emissions during atmospheric distillation arise from the combustion of fuels in the furnaces to heat the crude oil and fugitive emissions. Oily **sour water** (condensed steam containing hydrogen sulfate and ammonia) and oil is also generated in the fractionators.

Vacuum Distillation

Heavier fractions from the atmospheric distillation unit that cannot be distilled without cracking under its pressure and temperature conditions are vacuum distilled. Vacuum distillation is simply the distillation of petroleum fractions at a very low pressure (0. 2 to 0. 7 psia) to increase volatilization and separation. In most systems, the vacuum inside the fractionator is maintained with **steam ejectors**, vacuum pumps, and **barometric condensers**. The injection of superheated steam at the base of the vacuum fractionator column further reduces the partial pressure of the hydrocarbons in the tower, facilitating vaporization and separation. The heavier fractions from the vacuum distillation column are processed downstream into more valuable products through either cracking or coking operations.

Potential source of emissions from distillation of crude oil are the combustion of fuels in the furnace and some light gases leaving the top of the condensers on the vacuum distillation column. A certain amount of noncondensible light hydrocarbons and hydrogen sulfide pass through the condenser to a hot well, and then are discharged to the refinery sour fuel system or are vented to a process heater, flare or another control device to destroy hydrogen sulfide. The quantity of these emissions depends on the size of the unit, the type of feedstock, and the cooling water temperature. If barometric condensers are used in vacuum distillation, significant amounts of oily wastewater can be generated. Vacuum pumps and surface condensers have largely replaced barometric condensers in many refineries to eliminate this oily wastewater stream. Oily sour water is also generated in the fractionators.

Words and Expressions

desalt	v.	脱盐
reshape	v.	重塑
alkylation	n.	烷基化
vaporization	n.	汽化
condensation	n.	冷凝

corrosive	adj. 腐蚀性的
catalyst	n. 催化剂
deactivation	n. 减活,失活
demulsifier	n. 破乳剂
emulsion	n. 乳剂
potential	n. 电位
coalesce	v. 聚结
desalter	n. 脱盐器
sludge	n. 污泥,残渣
condense	v. 冷凝
methane	n. 甲烷
ethane	n. 乙烷

Technical Terms

heat exchanger	热交换器
atmospheric distillation	常压蒸馏
vacuum distillation	减压蒸馏
suspended solid	悬浮固体
distillation column	分馏塔
lighter fraction	轻质馏分
heavier fraction	重质馏分
side stream	侧线馏分
stripping tower	汽提塔
draw tray	抽出塔盘
hydrogen sulfide	硫化氢
sour gas/acid gas	酸性气体
sour water	含硫污水
steam ejector	蒸汽喷射泵
barometric condenser	大气冷凝器

Language Focus

1. For these reasons,no two refineries are alike.

此句为特指否定句,否定词"no"不否定整个句子,而是否定主语。"no"在这里可译为"没有",例如:No person of that name lives here. 这里没有叫

— 55 —

这个名字的人。

2. The desalting process also removes some of the metals and suspended solids which cause catalyst deactivation.

（参考译文为：脱盐过程中也能除去一些导致催化剂失活的金属和固体悬浮物。）

本句为限定性定语从句，"which"代替前面的 "some of the metals and suspended solids"。

3. The water used in crude desalting is often untreated or partially treated water from other refining process water.

此句的主谓结构是"the water... is... or water"，意为"未处理过的或部分处理过的"。

4. Another product produced in atmospheric distillation, as well as many other refinery processes, is the light, noncondensible refinery fuel gas (mainly methane and ethane).

［参考译文为：常压蒸馏和其他的炼油过程中的另一种产品是轻组分，即没有液化的气体燃料（主要是甲烷和乙烷）。］

此句中 as well as 连接两个名词词组，在此意为"除……之外"，"还"，例如：Women, as well as men, have a fundamental right to work. 除了男人，妇女也有工作的基本权利。

5. The sour gas is sent to the refinery sour gas treatment system which separates the fuel gas so that it can be used as fuel in the refinery heating furnaces.

（参考译文为：酸气送到炼油厂的酸气处理系统进行分离，以便在炼油厂加热设备中用作燃料。）

"which"引导的定语从句修饰先行词"the refinery sour gas treatment system"，"so that"引导目的状语从句，意为"以便"。

🔲 Reinforced Learning

Ⅰ. Answer the following questions for a comprehension of the text.

1. What processes do downstream processes and supporting operations include?

2. What are the steps to desalt?

3. How can we generate various fractions in distillation column at atmospheric pressure?

4. What are fractions obtained from atmospheric distillation?

5. Why should we use vacuum distillation?

II. Multiple choice: choose the correct one from the alternative answers to give the exact meaning of the word underlined.

1. The composition of crude oil can vary significantly depending on its source.

    A. beginning   B. cause   C. origin   D. geography

2. Refining crude oil into useful petroleum products can be separated into two phases and a number of supporting operations.

    A. secondary   B. showing   C. proving   D. assistant

3. Before separation into fractions, crude oil usually must first be treated to remove corrosive salts.

    A. move   B. lift   C. clean   D. eliminate

4. The desalting process creates an oily desalter sludge and a high temperature salt water waste stream which is typically added to other process wastewaters for treatment in the refinery wastewater treatment facilities.

    A. produces   B. gets   C. uses   D. invents

5. The desalted crude oil is then heated in a heat exchanger and furnace to about 750 degrees ($^\circ$F) and fed to a vertical, distillation column at atmospheric pressure where most of the feed is vaporized and separated into its various fractions by condensing on 30 to 50 fractionation trays, each corresponding to a different condensation temperature.

    A. material supplied to a machine

    B. the supplying of this machine

    C. the part of the machine supplying this material

    D. material getting out of the machine

6. Most of these can be sold as finished products, or blended with products from downstream processes.

    A. last   B. ended   C. produced   D. ruined

7. The steam strips the light-end components from the heavier components and both the steam and light-ends are fed back to the atmospheric distillation tower above the corresponding side stream draw tray.

    A. agreeing   B. equivalent   C. matching   D. related

8. Heavier fractions from the atmospheric distillation unit that cannot be distilled without cracking under its pressure and temperature conditions are vac-

uum distilled.

   A. part   B. equipment   C. department   D. section

9. The injection of superheated steam at the base of the vacuum fractionator column further reduces the partial pressure of the hydrocarbons in the tower, facilitating vaporization and separation.

   A. helping   B. quickening   C. assisting   D. simplifying

10. A certain amount of noncondensible light hydrocarbons and hydrogen sulfide pass through the condenser to a hot well, and then are discharged to the refinery sour fuel system or are vented to a process heater, flare or another control device to destroy hydrogen sulfide.

   A. fired   B. launched   C. shot   D. sent out

III. Multiple choice: read the four suggested translations and choose the best answer.

1. The desalted crude oil is then heated in a heat exchanger and furnace to about 750 degrees (℉) and fed to a vertical, distillation column at atmospheric pressure where most of the feed is vaporized and separated into its various fractions by condensing on 30 to 50 fractionation trays, each corresponding to a different condensation temperature.

   A. 根据   B. 回应   C. 与……一致   D. 与相关

2. These low–boiling point mixtures are in equilibrium with heavier components which must be removed.

   A. 和……相等   B. 和……平衡
   C. 和……一样   D. 和……均势

3. Air emissions during atmospheric distillation arise from the combustion of fuels in the furnaces to heat the crude oil and fugitive emissions.

   A. 自……升起   B. 由……引起
   C. 由于……增加   D. 因为……加剧

4. In most systems, the vacuum inside the fractionator is maintained with steam ejectors, vacuum pumps, and barometric condensers.

   A. 用……维持   B. 使……继续
   C. 用……保护   D. 用……保养

5. Vacuum pumps and surface condensers have largely replaced barometric condensers in many refineries to eliminate this oily wastewater stream.

   A. 大体上   B. 基本上   C. 多半   D. 主要地

Ⅳ. Put the following sentences into Chinese.

1. The composition of crude oil can vary significantly depending on its source.

2. One of the most important operations in a refinery is the initial distillation of the crude oil into its various boiling point fractions.

3. The side streams are each sent to a different small stripping tower containing four to 10 trays with steam injected under the bottom tray.

4. In most systems, the vacuum inside the fractionator is maintained with steam ejectors, vacuum pumps, and barometric condensers.

5. If barometric condensers are used in vacuum distillation, significant amounts of oily wastewater can be generated.

Ⅴ. Put the following paragraphs into Chinese.

1. Petroleum refineries are a complex system of multiple operations and the operations used at a given refinery depend upon the properties of the crude oil to be refined and the desired products. For these reasons, no two refineries are alike. Portions of the outputs from some processes are refed back into the same process, fed to new processes, fed back to a previous process, or blended with other outputs to form finished products.

2. Before separation into fractions, crude oil usually must first be treated to remove **corrosive** salts. The desalting process also removes some of the metals and suspended solids which cause catalyst deactivation. Desalting involves the mixing of heated crude oil with water so that the salts are dissolved in the water. The water must then be separated from the crude oil in a separating vessel by adding demulsifier chemicals to assist in breaking the emulsion and/or, more commonly, by applying a high potential electric field across the settling vessel to coalesce the polar salt water droplets.

2.4　Downstream Processing

Guidance to Reading

Downstream processes change the molecular structure of hydrocarbon molecules either by breaking them into smaller molecules, joining them to form larger molecules, or reshaping them into higher quality molecules. The processing methods include thermal cracking, coking, catalytic cracking, catalytic hydro-cracking and so on. At present, thermal cracking has been largely replaced by catalytic cracking while "fluid coking" is expected to be an important process in the future instead of "delayed coking".

Text

Certain fractions from the distillation of crude oil are further refined in **thermal cracking** (**visbreaking**), coking, catalytic cracking, **catalytic hydrocracking**, **hydrotreating**, alkylation, **isomerization**, **polymerization**, **catalytic reforming**, **solvent extraction**, **dewaxing**, propane deasphalting and other operations. These downstream processes change the molecular structure of hydrocarbon molecules either by breaking them into smaller molecules, joining them to form larger molecules, or reshaping them into higher quality molecules. For many of the operations discussed below, a number of different techniques are used in the industry.

1. Thermal Cracking/Visbreaking

Thermal cracking, or visbreaking, uses heat and pressure to break large hydrocarbon molecules into smaller, lighter molecules. The process has been largely replaced by catalytic cracking and some refineries no longer employ thermal cracking. Both processes reduce the production of less valuable products such as heavy fuel oil, and increase the feed stock to the catalytic cracker and gasoline yields. In thermal cracking, heavy gas oils and residue from the vacuum distillation process are typically the feed.

2. Coking

Coking is a cracking process used primarily to reduce refinery production of low – value residual fuel oils to transportation fuels, such as gasoline and diesel. As part of the upgrading process, coking also produces petroleum coke, which is essentially solid carbon with varying amounts of impurities, and is used as a fuel for power plants if the sulfur content is low enough. Coke also has nonfuel applications as a raw material for many carbon and **graphite** products including **anodes** for the production of aluminum, and furnace **electrodes** for the production of elemental phosphorus, **titanium dioxide**, calcium carbide and silicon carbide. A number of different processes are used to produce coke; "**delayed coking**" is the most widely used today, but "**fluid coking**" is expected to be an important process in the future. Fluid coking produces a higher grade of coke which is increasingly in demand. In delayed coking operations, the same basic process as thermal cracking is used except feed streams are allowed to react longer without being cooled. The delayed coking feed stream of residual oils from various upstream processes is first introduced to a fractionating tower where residual lighter materials are drawn off and the heavy ends are condensed. The heavy ends are removed and heated in a furnace to about 900 ~ 1,000 degrees (°F) and then fed to an **insulated** vessel called a **coke drum** where the coke is formed. When the coke drum is filled with product, the feed is switched to an empty parallel drum. Hot vapors from the coke drums, containing cracked lighter hydrocarbon products, hydrogen sulfide, and ammonia, are fed back to the fractionator where they can be treated in the sour gas treatment system or drawn off as intermediate products. Steam is then injected into the full coke drum to remove hydrocarbon vapors, water is injected to cool the coke, and the coke is removed. Typically, **high pressure water jets** are used to cut the coke from the drum.

Air emissions from coking operations include the process heater flue gas emissions, fugitive emissions and emissions that may arise from the removal of the coke from the coke drum. The injected steam is condensed and the remaining vapors are typically flared. Wastewater is generated from the coke removal and cooling operations and from the steam injection. In addition, the removal of coke from the drum can release particulate emissions and any remaining hydrocarbons to the atmosphere.

3. Catalytic Cracking

Catalytic cracking uses heat, pressure and a catalyst to break larger hydrocarbon molecules into smaller, lighter molecules. Catalytic cracking has largely replaced thermal cracking because it is able to produce more gasoline with a higher octane and less heavy fuel oils and light gases. Feed stocks are light and heavy oils from the crude oil distillation unit which are processed primarily into gasoline as well as some fuel oil and light gases. Most catalysts used in catalytic cracking consist of mixtures of **crystalline** synthetic **silica – alumina**, termed "**zeolites**", and **amorphous** synthetic silica – alumina. The catalytic cracking processes, as well as most other refinery catalytic processes, produce coke which collects on the catalyst surface and diminishes its catalytic properties. The catalyst, therefore, needs to be regenerated continuously or periodically essentially by burning the coke off the catalyst at high temperatures. The method and frequency in which catalysts are regenerated are a major factor in the design of catalytic cracking units. A number of different catalytic cracking designs are currently in use in the U. S. , including fixed – bed reactors, moving – bed reactors, fluidized – bed reactors, and once – through units.

In the fluidized – bed process, oil and oil vapor preheated to 500 to 800 degrees (℉) is contacted with hot catalyst at about 1,300 (℉) either in the reactor itself or in the feed line (riser) to the reactor. The catalyst is in a fine, **granular** form which, when mixed with the vapor, has many of the properties of a fluid. The fluidized catalyst and the reacted hydrocarbon vapor separate mechanically in the reactor and any oil remaining on the catalyst is removed by steam **stripping**. The cracked oil vapors are then fed to a fractionation tower where the various desired fractions are separated and collected. The catalyst flows into a separate vessel(s) for either single-or two-stage regeneration by burning off the coke **deposits** with air.

In the moving – bed process, oil is heated to up to 1,300 degrees (℉) and is passed under pressure through the reactor where it comes into contact with a catalyst flow in the form of **beads** or **pellets**. The cracked products then flow to a fractionating tower where the various compounds are separated and collected. The catalyst is regenerated in a continuous process where deposits of coke on the catalyst are burned off. Some units also use steam to strip remaining hydrocarbons and oxygen from the catalyst before being fed back to the oil stream.

4. Catalytic Hydrocracking

Catalytic hydrocracking normally utilizes a fixed-bed catalytic cracking reactor with cracking occurring under substantial pressure (1,200 to 2,000 psig) in the presence of hydrogen. Feedstocks to hydrocracking units are often those fractions that are the most difficult to crack and cannot be cracked effectively in catalytic cracking units. These include: middle distillates, cycle oils, residual fuel oils and **reduced crudes**. The hydrogen suppresses the formation of heavy residual material and increases the yield of gasoline by reacting with the cracked products. However, this process also breaks the heavy, sulfur and nitrogen bearing hydrocarbons and releases these impurities to where they could potentially foul the catalyst. For this reason, the feedstock is often first hydrotreated to remove impurities before being sent to the catalytic hydrocracker. Sometimes hydrotreating is accomplished by using the first reactor of the hydrocracking process to remove impurities. Water also has a detrimental effect on some hydrocracking catalysts and must be removed before being fed to the reactor. The water is removed by passing the feed stream through a silica gel or molecular **sieve dryer**. Depending on the products desired and the size of the unit, catalytic hydrocracking is conducted in either single stage or multi – stage reactor processes. Most catalysts consist of a crystalline mixture of silica – alumina with small amounts of **rare earth** metals.

Hydrocracking feedstocks are usually first hydrotreated to remove the hydrogen sulfide and ammonia that will poison the catalyst. Sour gas and sour water streams are produced at the fractionators; however, if the hydrocracking feedstocks are first hydrotreated to remove impurities, both streams will contain relatively low levels of hydrogen sulfide and ammonia. Hydrocracking catalysts are typically regenerated off – site after two to four years of operation. Therefore, little or no emissions are generated from the regeneration processes. Air emissions arise from the process heater, vents, and fugitive emission.

🔲 Words and Expressions

visbreaking	n. 减黏裂化;降黏
hydrotreat	v. 加氢处理
isomerization	n. 异构化
polymerization	n. 聚合
dewax	v. 脱蜡
graphite	n. 石墨
anode	n. 阳极
electrode	n. 电极
insulated	adj. 绝热的
crystalline	adj. 结晶的
silica-alumina	硅—铝
zeolite	n. 沸石
amorphous	adj. 非晶体的
granular	adj. 粒状的
strip	v. 汽提

deposit	n. 沉积物
bead	n. 珠子
pellet	n. 颗粒

🏠 Technical Terms

thermal cracking	热裂解
catalytic hydrocrack	加氢催化裂化
catalytic reform	催化重整
solvent extraction	溶剂萃取
titanium dioxide	钛白粉
delayed coke	延迟焦化
fluid coke	流化焦化
coke drum	焦化塔
high pressure water jet	高压水射流
reduced crude	拔顶油
sieve dryer	筛网烘干机

🏠 Language Focus

1. These downstream processes change the molecular structure of hydrocarbon molecules either by breaking them into smaller molecules, joining them to form larger molecules, or reshaping them into higher quality molecules.

（参考译文为：这些下游加工过程可以将烃类化合物裂化为小分子化合物或聚合为大分子化合物，也可以将其重组转化为更高分子量的化合物，从而改变烃类化合物的分子结构。）

此句中包含了"either A, B or C"的结构，表示选择关系，任意一个均可。

2. Hot vapors from the coke drums, containing cracked lighter hydrocarbon products, hydrogen sulfide, and ammonia, are fed back to the fractionators where they can be treated in the sour gas treatment system or drawn off as intermediate products.

（参考译文为：从焦炭出来的热蒸汽含有较轻的烃类产品，例如硫化氢和氨，这些热蒸汽要返回到分馏塔由酸气处理系统处理，或作为中间产品取出。）

本句主干为"Hot vapors are fed back to the fractionators"，意为热蒸汽出来后重新进入分馏塔。"where"引导的定语从句修饰先行词"fractionators"。

3. The catalyst is in a fine, granular form which, when mixed with the va-

por, has many of the properties of a fluid.

该句中"which has many of the properties of a fluid"做"form"的定语，when mixed with the vapor 为插入的时间状语。"granular"一词意为"粒状的"。

4. The catalyst flows into a separate vessel(s) for either single-or two-stage regeneration by burning off the coke deposits with air.

（参考译文为：催化剂流入另一个容器中，在空气中燃烧掉积炭，实现一次或两次催化剂再生。）

本句中介词的使用不仅可以简化句子，也能实现句子言简意赅、准确达意的功能，在科技英语中常用。

5. The catalyst is regenerated in a continuous process where deposits of coke on the catalyst are burned off.

（参考译文为：只要催化剂积炭燃烧掉，催化剂便可连续再生。）

本句中"where"引导定语从句，修饰"a continuous process"。

Reinforced Learning

I. Answer the following questions for a comprehension of the text.

1. What operations will certain fractions from the distillation of crude oil be further refined in?

2. What's the major difference between delay cooking and thermal coking?

3. How to generate the coke in delay coking?

4. What are the forms of air emissions in coking?

5. How to regenerate catalyst in fluidized – bed process and moving – bed process?

II. Multiple choice: choose the correct one from the alternative answers to give the exact meaning of the word underlined.

1. These downstream processes change the molecular structure of hydrocarbon molecules either by breaking them into smaller molecules, joining them to form larger molecules, or reshaping them into higher quality molecules.

　　A. sticking　　　B. connecting　　　C. combining　　　D. uniting

2. Thermal cracking, or visbreaking, uses heat and pressure to break large hydrocarbon molecules into smaller, lighter molecules.

　　A. separate　　　B. disperse　　　C. shorten　　　D. exchange

3. Coke also has nonfuel <u>applications</u> as a raw material for many carbon and graphite products including anodes for the production of aluminum, and furnace electrodes for the production of elemental phosphorus, titanium dioxide, calcium carbide and silicon carbide.

    A. requests   B. technology   C. use   D. effects

4. Typically, high pressure water jets are used to <u>cut</u> the coke from the drum.

    A. break   B. resolve   C. separate   D. reduce

5. In addition, the <u>removal</u> of coke from the drum can release particulate emissions and any remaining hydrocarbons to the atmosphere.

    A. loose   B. exemption   C. unfasten   D. free

6. The catalytic cracking processes, as well as most other refinery catalytic processes, produce coke which <u>collects</u> on the catalyst surface and diminishes its catalytic properties.

    A. cumulates   B. generates   C. produces   D. falls

7. The catalyst, therefore, needs to be <u>regenerated</u> continuously or periodically essentially by burning the coke off the catalyst at high temperatures.

    A. re – established   B. renewed

    C. cleaned   D. rejuvenated

8. In the fluidized – bed process, oil and oil vapor preheated to 500 to 800 degrees (℉) is contacted with hot catalyst at about 1,300 (℉) either in the reactor itself or in the feed <u>line</u> (riser) to the reactor.

    A. path   B. tunnel   C. process   D. time

9. The hydrogen <u>suppresses</u> the formation of heavy residual material and increases the yield of gasoline by reacting with the cracked products.

    A. bans   B. controls   C. prevents   D. subdues

10. However, this process also breaks the heavy, sulfur and nitrogen <u>bearing</u> hydrocarbons and releases these impurities to where they could potentially foul the catalyst.

    A. included   B. suffered   C. owning   D. enduring

III. Multiple choice: read the four suggested translations and choose the best answer.

1. The process has been largely replaced by catalytic cracking and some refineries <u>no longer</u> employ thermal cracking.

    A. 不再   B. 减少(次数)   C. 降低(程度)   D. 不多于

2. The delayed coking feed stream of residual oils from various upstream processes is first introduced to a fractionating tower where residual lighter materials are drawn off and the heavy ends are condensed.

　　A. 取出　　　B. 拿出　　　C. 释放　　　D. 发散

3. When the coke drum is filled with product, the feed is switched to an empty parallel drum.

　　A. 转移到　　B. 打开　　　C. 拐弯到　　　D. 连接至

4. In the fluidized – bed process, oil and oil vapor preheated to 500 to 800 degrees (℉) is contacted with hot catalyst at about 1,300 (℉) either in the reactor itself or in the feed line (riser) to the reactor.

　　A. 连接　　　B. 转移　　　C. 接触　　　D. 交换

5. Sometimes hydrotreating is accomplished by using the first reactor of the hydrocracking process to remove impurities.

　　A. 由……完成　　　　　B. 由……帮助
　　C. 由……做出成绩　　　D. 因……获得成就

Ⅳ. Put the following sentences into Chinese.

1. In thermal cracking, heavy gas oils and residue from the vacuum distillation process are typically the feed.

2. When the coke drum is filled with product, the feed is switched to an empty parallel drum.

3. The injected steam is condensed and the remaining vapors are typically flared.

4. Catalytic cracking uses heat, pressure and a catalyst to break larger hydrocarbon molecules into smaller, lighter molecules.

5. The catalyst, therefore, needs to be regenerated continuously or periodically essentially by burning the coke off the catalyst at high temperatures.

V. Put the following paragraphs into Chinese.

1. Thermal cracking, or visbreaking, uses heat and pressure to break large hydrocarbon molecules into smaller, lighter molecules. The process has been largely replaced by catalytic cracking and some refineries no longer employ thermal cracking. Both processes reduce the production of less valuable products such as heavy fuel oil, and increase the feed stock to the catalytic cracker and gasoline yields. In thermal cracking, heavy gas oils and residue from the vacuum

distillation process are typically the feed.

2. Hydrocracking feedstocks are usually first hydrotreated to remove the hydrogen sulfide and ammonia that will poison the catalyst. Sour gas and sour water streams are produced at the fractionators; however, if the hydrocracking feedstocks are first hydrotreated to remove impurities, both streams will contain relatively low levels of hydrogen sulfide and ammonia. Hydrocracking catalysts are typically regenerated off – site after two to four years of operation. Therefore, little or no emissions are generated from the regeneration processes. Air emissions arise from the process heater, vents, and fugitive emission.

Chapter 3　Primary Refining

3. 1　Distilling

⌸ Guidance to Reading

*Petroleum refining begins with primary physical separation, or the **distillation**, or fractionation, of crude oils into separate hydrocarbon groups. Crude oil is separated in atmospheric and vacuum distillation towers, into "fractions" or "cuts" with various boiling – point ranges. Distilling columns and cracking towers are quite different. Knowledge of distilling columns structure and working mechanisms is essential for one to understand primary refining process.*

⌸ Text

A casual passerby of a refinery can make an easy mistake by referring to the many tall **columns** inside as "**cracking towers**". In fact, most of them are distilling columns of one sort or another. Cracking towers are usually shorter and **squatter**.

Distilling units are the clever invention of process engineers who **exploit** the important characteristics of the distillation curve. The **mechanism** they use is not too complicated. However, in the interest of completeness and familiarity, you can cover the **rudiment**s here.

The Simple Still

For years Kentucky **moonshiners** used the simple still in Fig. 3. 1 to separate the white lightning from the **dregs**. After the sour mash **ferment**ed – i. e. , a portion of it had slowly undergone a chemical change to alcohol—they heated it to the boiling range of the alcohol. The white lightning **vaporized** as a vapor, it is less dense (lighter) than liquid. It moved out of the liquid, then through the **condenser** where it cooled and turned back to liquid. The liquid left in the still was **discarded**. The liquid that ended up in the condenser was bottled. A process engineer would call this a simple batch process distillation.

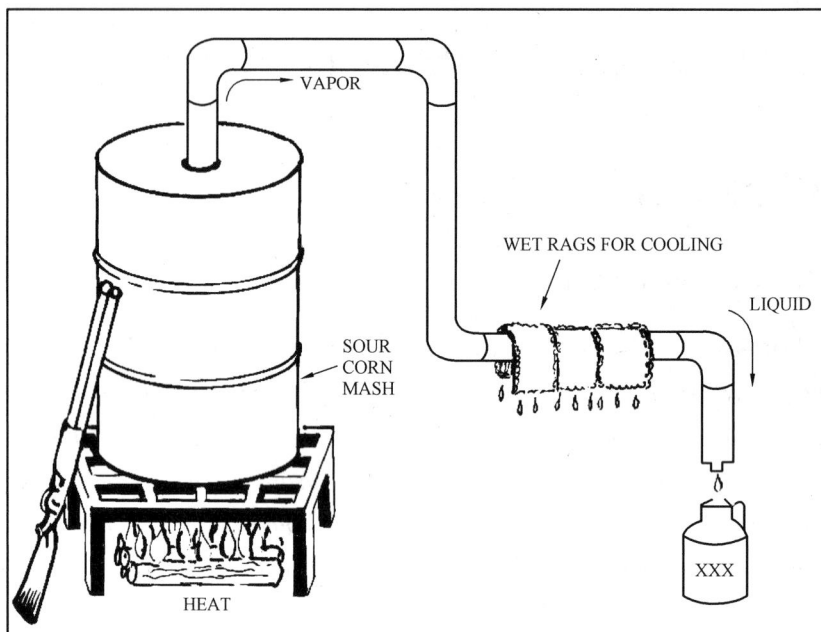

Fig. 3. 1   The Moonshiner's Still

If the moonshiners wanted to sell a better – than – average product, they might have run the product through a second batch still much like the first. There they could have separated the best part of the liquor from some of the **non – alcoholic impurities** that inevitably flowed along with the overhead in the first still. Some impurities might have gone overhead because of the inadequacy of the temperature controls or because the moonshiners wanted to be sure they got all they could so they set the temperature a little high on the first batch.

This two – step operation could be made into the continuous operation shown in Fig. 3. 2. In fact, many early oil distilling operations looked like that.

The Distilling Column

The two – step batch distilling operation is obviously not suited for handling several hundred thousand barrels per day of crude oil with five or six different components being separated. A distilling column can do it on a continuous basis with fewer facilities and much less labor and energy consumption.

Fig. 3. 3 shows from afar what happens at a crude distilling column. Crude goes in, and the products go out—gases (butane and lighter) , gasoline, naphtha, kerosene, **light gas oil**, **heavy gas oil**, and **residue**.

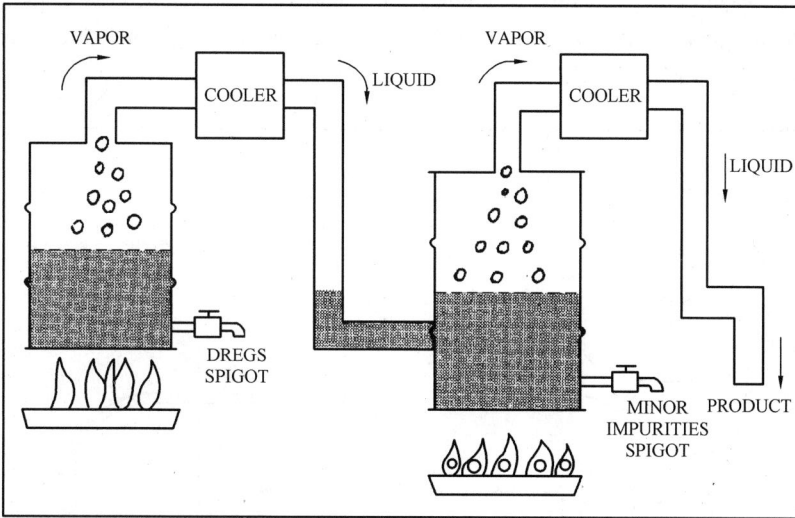

Fig. 3. 2　Two Stage Batch Still

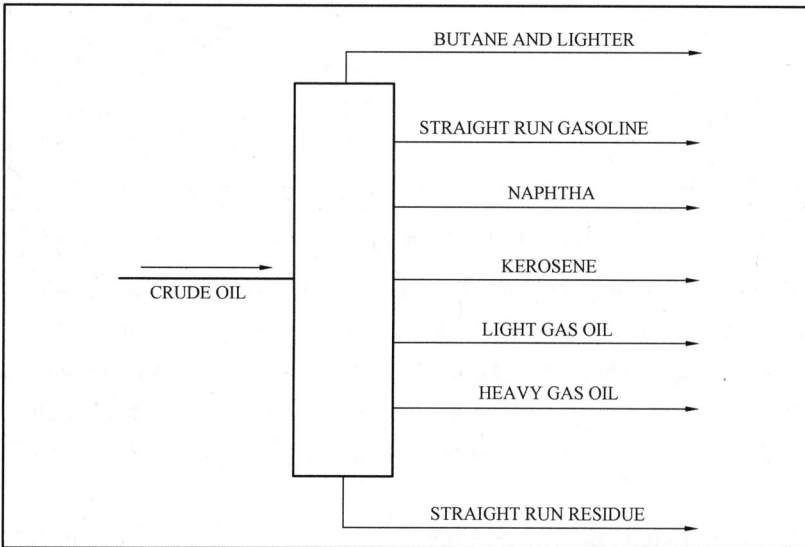

Fig. 3. 3　Distilling

What goes on inside the distilling column is more complicated. The first piece of equipment important to the operation, the **charge pump**, moves the crude from the storage tank through the system (Fig. 3. 4). The crude is first pumped through a **furnace** where it is heated to a temperature of about 750 ℉.

More than half of the crude oil changes to vapor form as the furnace heats it to this temperature. This **combination** of liquid and vapor is then introduced to the distilling column.

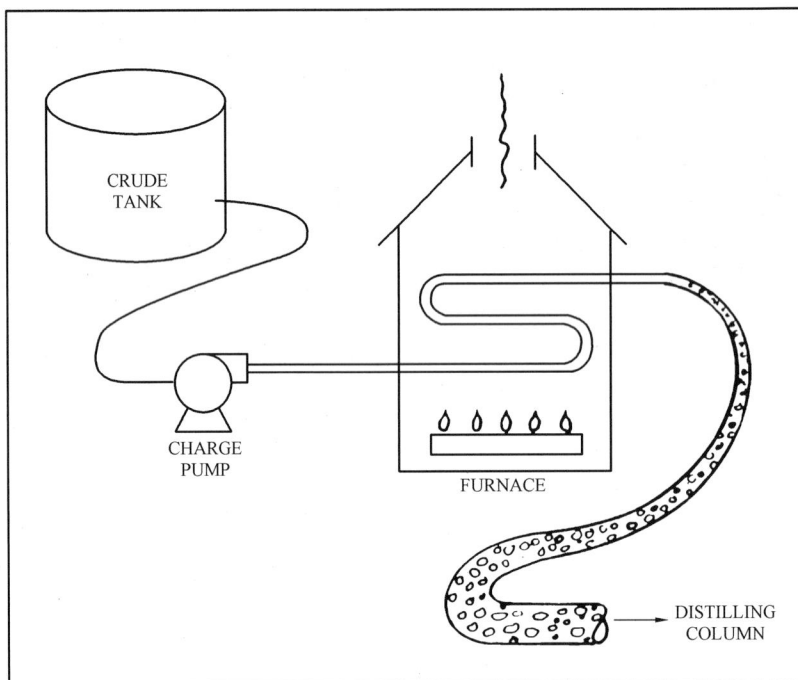

Fig. 3. 4　Crude Oil Feed to Distilling

　　Inside the distilling column there is a set of **trays** with **perforations** in them. The perforations permit the vapors to rise through the column. When the crude liquid/vapor charge hits the inside of the distilling column gravity causes the denser (heavier) liquid to drop toward the column bottom, but the less dense(lighter)vapors start moving through the trays toward the top, as Fig. 3. 5 shows.

　　The perforations in the trays are fitted with a device called bubble caps (Fig. 3. 6). Their purpose is to force the vapor coming up through the trays to bubble through the liquid standing several inches deep on that tray. This bubbling is the **essence** of the distilling operation: the hot vapor (starting at 750 °F) bubbles through the liquid. Heat transfers from the vapor to the liquid during the bubbling. As the vapor bubbles cool a little, some of the hydrocarbons in them will change from the vapor to the liquid state; the temperature of the vapor

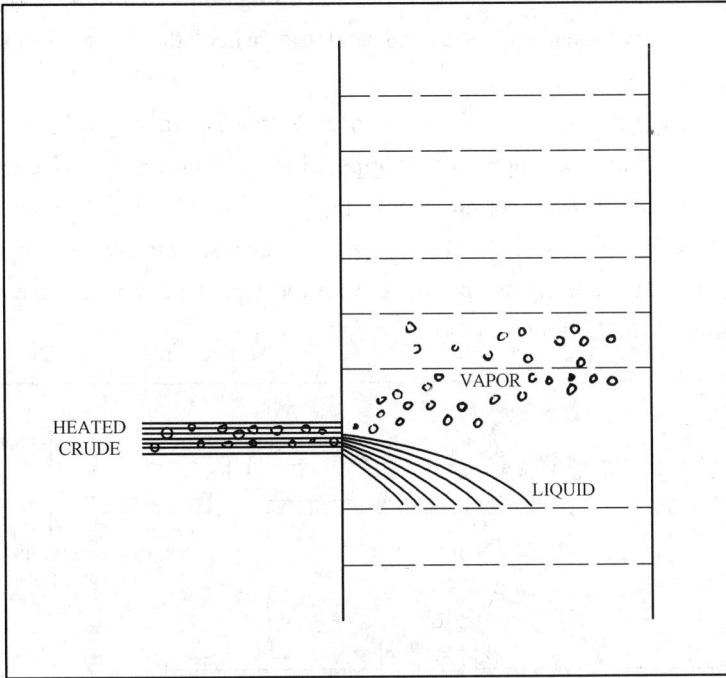

Fig. 3. 5　Crude Entering the Distilling Column

drops and the lower temperature of the liquid causes whatever heavier compounds remain in the vapor to condense(liquefy).

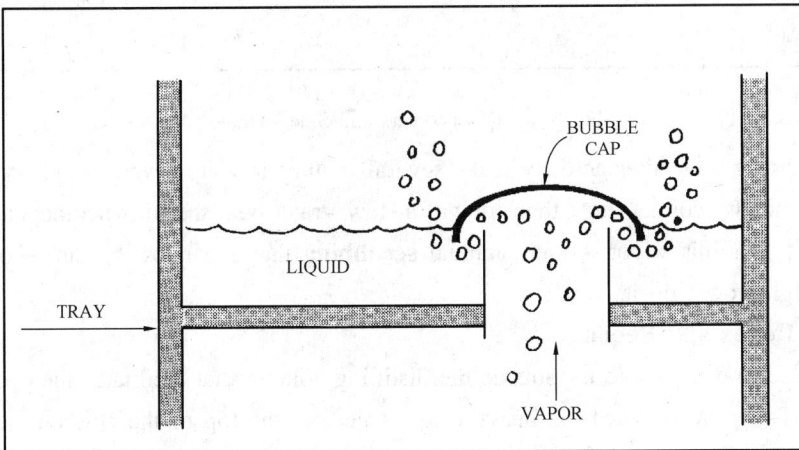

Fig. 3. 6　Bubble Cap on a Distilling Column Tray

After passing through the liquid and **shedding** some of the heavier hydrocarbons, the vapor then moves to the next tray where the same process takes place.

Meanwhile, the amount of liquid on each tray is growing as some of the hydrocarbons from the vapor are **stripped**. Fig. 3. 7 shows a device called a **downcomer**, installed to permit excess liquid to overflow to the next lower tray. At several levels on the column, the sidedraws shown take the liquid distilled product off—the lighter products from the top of the column, the heavier liquids toward the bottom.

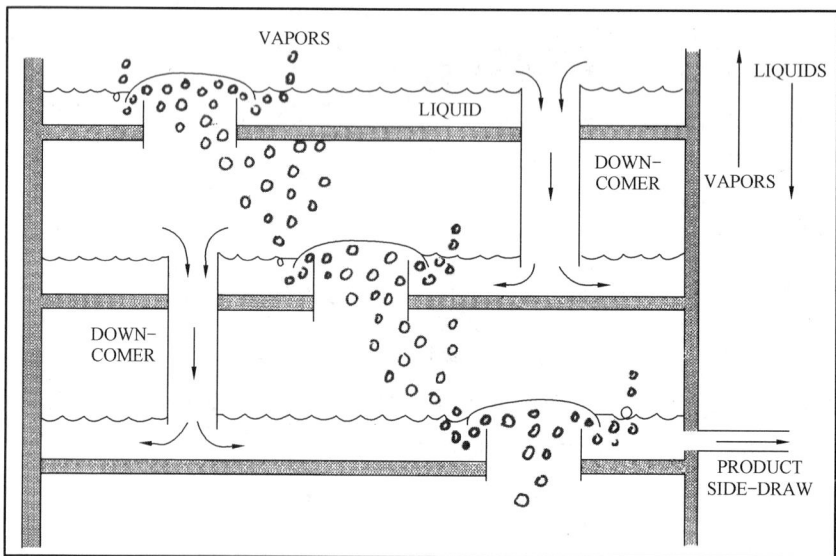

Fig. 3. 7   Downcomers and Side – Draws

Some molecules actually make several round trips: up several trays as vapor, finally condensing, then down a few trays via the downcomer as a liquid. It's this vapor – liquid mutual **scrubbing** that separates the cuts – once through won't do it.

Reflux and Reboil

Several things go on outside the distilling column that facilitates the operation. To assure some of the heavies don't get out the top of the column, occasionally some of the vapor will be run through a cooler. Whatever is condensed is reintroduced to a lower tray. Remaining vapor is sent off as product. The process is a form of refluxing(Fig. 3. 8).

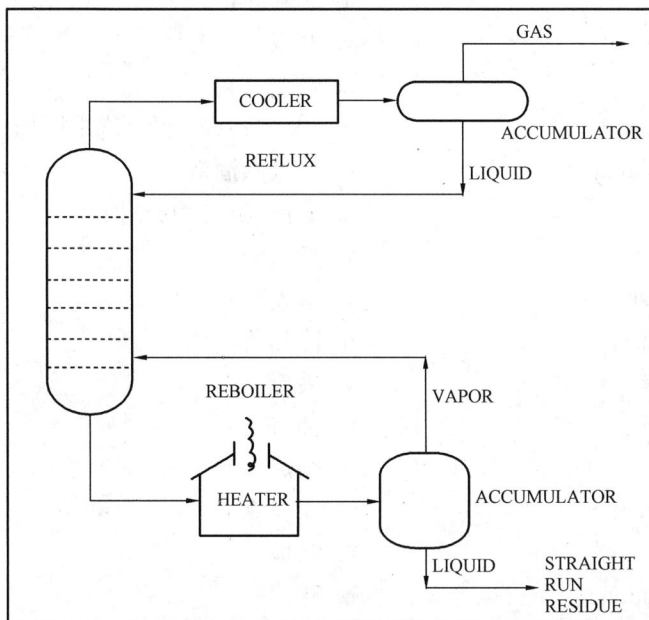

Fig. 3. 8 Reboil and Reflux

Conversely, some of the light hydrocarbon could be entrained on the bottom of the column where the liquid part of the crude oil ended. So a **sidedraw** may be used to recirculate the liquid through a heater to drive off any lighter hydrocarbons for reintroduction as a vapor at some lower level in the distilling column. This is called reboiling.

Cut Points

For analyzing and controlling distilling operations, the key **parameters** are cut points, the temperatures at which the various distilling products are separated. The temperature at which a product (or cut or fraction) begins to boil is called the initial boiling point (IBP). The temperature at which it is 100% vaporized is the end point (EP). So every cut has two cut points. The end point naphtha is the IBP of kerosene. At a cut point, the EP and IBP of the two adjacent cuts are nominally the same.

Words and Expressions

distill	v.	蒸馏
column	n.	柱体,柱状物

squat	a. 矮胖的
exploit	v. 开发
mechanism	n. 机制
rudiment	n. 初步,初级（pl.）基本原理
moonshiner	n. 私人酿酒者
dreg	n. 渣滓;沉淀物;糟粕
ferment	v. 使发酵
vaporize	v. 蒸发
condenser	n. 冷却器
discard	v. 丢弃
residue	n. 油渣
furnace	n. 熔炉
combination	n. 混合物
tray	n. 筛板
perforation	n. 孔
essence	n. 本质
shed	v. 使脱落
strip	v. 除去
downcomer	n. 降液管
scrub	v. 洗涤
sidedraw	n. 侧线
parameter	n. 参数

Technical Terms

cracking tower	裂化塔
non – alcoholic impurities	非酒精杂质
light gas oil	轻瓦斯油
heavy gas oil	重瓦斯油
charge pump	加料泵

Language Focus

1. Distilling units are the clever invention of process engineers who exploit the important characteristics of the distillation curve.

本句中"who"引导的定语从句"who exploit the important characteristics

of the distillation curve"修饰先行词"process engineers"。句中"exploit"一词在科技英语中常用,通常表示"深入研究"或"充分利用"。

2. If the moonshiners wanted to sell a better – than – average product, they might have run the product through a second batch still much like the first.

本句中"better – than – average"指的是高于一般水平的。另"might have done"表示"可能性",酿酒师们可能会按照第一次蒸馏的方法将产品进行二次蒸馏。

3. There they could have separated the best part of the liquor from some of the non – alcoholic impurities that inevitably flow along with the overhead in the first still.

本句中"could have done"表示可以做某事。另外"that inevitably flowed along with the overhead in the first still"意为"这些杂质是在第一次蒸馏中不可避免地流向蒸馏塔塔顶的部分"。

4. The two – step batch distilling operation is obviously not suited for handling several hundred thousand barrels per day of crude oil with five or six different components being separated.

本句中"with five or six different components being separated"为独立主格,意为"而且原油中含有 5~6 种组分需要分离"。with + 名词 + 分词(形容词、副词或不定式)的结构是独立主格的主要结构,例如:With winter coming on, it's time to buy warm clothes. 随着冬天的到来,该买厚衣服了。

5. The crude is first pumped through a furnace where it is heated to a temperature of about 750 ℉.

本句中"where it is heated to a temperature of about 750 ℉"为"where"引导的定语从句修饰"a furnace"。根据汉语的习惯,定语应翻译在先行词的前面:原油首先进入一个加热到 750 ℉的熔炉中。

6. After passing through the liquid and shedding some of the heavier hydrocarbons, the vapor then moves to the next tray where the same process takes place.

本句中"passing through the liquid 和 shedding some of the heavier hydrocarbons"为"and"连接的并列结构,表示顺序关系,做时间状语。句中"where the same process takes place"为定语从句修饰先行词"tray"。

7. It's this vapor – liquid mutual scrubbing that separates the cuts – once through won't do it.

本句为"it"引导的强调句,强调句中通常将最重要的成分放在"it"后面,再接"that"从句。"that"是强调句的主要标志之一。例如:It was an old beggar that Mary met in the street yesterday. 玛丽昨天在路上看到的是个老乞丐。

8. The temperature at which a product (or cut or fraction) begins to boil is called the initial boiling point (IBP).

本句中"at which a product (or cut or fraction) begins to boil"为定语从句修饰"temperature",介词提前于关系代词是定语从句中常见的形式,例如: This is the house in which I used to live.

Reinforced Learning

I. Answer the following questions for a comprehension of the text.

1. What is the wrong conception a casual passerby of a refinery usually hold when they saw many tall columns inside?

2. How did the moonshiners use to do to separate the white lightning from the dregs?

3. Give a brief introduction to what goes on inside the distilling column.

4. What is the purpose of refluxing?

5. What are the key parameters for analyzing and controlling operations?

II. Multiple choice: choose the correct one from the alternative answers to give the exact meaning of the word underlined.

1. A casual passerby of a refinery can make an easy mistake by referring to the many tall columns inside as "cracking towers".

    A. random   B. careless   C. incidental   D. irregular

2. Distilling units are the clever invention of process engineers who exploit the important characteristics of the distillation curve.

    A. develop   B. study   C. explore   D. achieve

3. The liquid left in the still was discarded.

    A. scrapped   B. shed   C. recycled   D. processed

4. There they could have separated the best part of the liquor from some of the non–alcoholic impurities that inevitably followed along with the overhead in the first still.

    A. substances   B. dopants   C. constituents   D. matter

5. Meanwhile, the amount of liquid on each tray is growing as some of the

hydrocarbons from the vapor are stripped.

A. cut   B. undressed   C. removed   D. extracted

6. To assure some of the heavies don't get out the top of the column, occasionally some of the vapor will be run through a cooler.

A. incidentally   B. exceptionally   C. instantly   D. sometimes

7. For analyzing and controlling distilling operations, the key parameters are cut points, the temperatures at which the various distilling products are separated.

A. boundaries   B. factors   C. limits   D. constraints

8. So every cut has two cut points. At a cut point, the EP and IBP of the two adjacent cuts are nominally the same.

A. frequently   B. regularly   C. usually   D. customarily

III. Multiple choice: read the four suggested translations and choose the best answer.

1. The white lightning vaporized as a vapor, it is less dense (lighter) than liquid.

A. 气化   B. 挥发   C. 液化   D. 冷凝

2. A distilling column can do it on a continuous basis with fewer facilities and much less labor and energy consumption.

A. 促进   B. 设备   C. 连接器   D. 容器

3. Conversely, some of the light hydrocarbon could be entrained on the bottom of the column where the liquid part of the crude oil ended.

A. 相应地说   B. 相对而言   C. 相反地   D. 相互地

4. So a sidedraw may be used to recirculate the liquid through a heater to drive off any lighter hydrocarbons for reintroduction as a vapor at some lower level in the distilling column.

A. 再返回   B. 再循环   C. 再介绍   D. 再吸收

5. The temperature at which a product (or cut or fraction) begins to boil is called the initial boiling point (IBP).

A. 最终的   B. 最底的   C. 最高的   D. 最初的

IV. Put the following sentences into Chinese.

1. Distilling units are the clever invention of process engineers who exploit the important characteristics of the distillation curve.

2. The two – step batch distilling operation is obviously not suited for handling several hundred thousand barrels per day of crude oil with five or six different components being separated.

3. Crude goes in, and the products go out—gases (butane and lighter), gasoline, naphtha, kerosene, light gas oil, heavy gas oil, and residue.

4. Meanwhile, the amount of liquid on each tray is growing as some of the hydrocarbons from the vapor are stripped.

5. Conversely, some of the light hydrocarbon could be entrained on the bottom of the column where the liquid part of the crude oil ended.

V. Put the following paragraphs into Chinese.

1. Some molecules actually make several round trips: up several trays as vapor, finally condensing, then down a few trays via the downcomer as a liquid. It's this vapor – liquid mutual scrubbing that separates the cuts—once through won't do it.

2. Conversely, some of the light hydrocarbon could be entrained on the bottom of the column where the liquid part of the crude oil ended. So a sidedraw may be used to recirculate the liquid through a heater to drive off any lighter hydrocarbons for reintroduction as a vapor at some lower level in the distilling column. This is called reboiling.

3. 2　Vacuum Flashing

⌐ Guidance to Reading

In the distillation process, all the uncontrolled symptoms of cracking could occur. Temperature sensitive materials also require vacuum distillation without damaging the product. Vacuum distillation is referred to as low temperature distillation, usually at less than atmospheric pressure. To avoid them, refiners developed **vacuum flashing***.*

⌐ Text

The discussions so far about distillation curves and distilling columns have treated the shape of the curves at temperatures close to 900 °F somewhat **vaguely**, but on purpose. There is a phenomenon that happens at these temperatures called **cracking**. Refiners use cracking **extensively**, but in controlled ways.

Effects of Low Pressure

Suppose you own a chain of restaurants that **specialize in** serving baked potatoes—one in Houston, Texas and one in Denver, Colorado. Do you know that baking potatoes in Denver takes 20 minutes longer than in Houston? (I found that out the hard way when I went skiing in the Rocky Mountains and cooked my own supper one night).

The reason is that at the higher **altitude** of Denver, the **atmospheric pressure** is lower. When people say the air in the mountains is thinner, they really mean there is less air per cubic inch. Since the pressure is lower, water will boil at a lower temperature. Even though you have the oven set at 375 °F, the boiling moisture will keep the potato at the lower temperature of 205 °F in Denver, not the 212 °F in Houston. Thus the potato doesn't ever get as hot in Denver as it does in Houston and it takes longer to cook.

All this **illustrates** the simple relationship between pressure and temperature. The process of heating lets the **molecules** absorb enough energy to escape from the liquid form to the gaseous form. The rate at which they escape depends on the rate at which the heat is delivered (how high you turn the burner) and the pressure of the air above them. The lower the pressure, the less energy has to be **transmitted**, and the lower the temperature at which the vapor will start forming in the liquid, i. e. boil.

The point is, the lower the pressure, the lower the boiling temperature.

Vacuum Flashing

Apply the pressure/boiling relationship to the crude oil cracking problem. The straight run **residue** will crack if the temperature goes too high, but straight run residue needs to be separated into more cuts. The solution is to do the distillation at reduced pressure.

In the vacuum flasher shown in Fig. 3. 9, straight run residue from the distilling unit, while it is still hot, is pumped to the flasher. The flasher is a large diameter vessel, too **squat** to be called a column, but that's more or less what it is. The straight run residue leaves the pump and heads down the line and into the flasher, where the pressure has been lowered well below atmospheric pressure. At that low pressure the lighter portions of the straight run residue will vaporize, but at that temperature, no cracking will take place.

Various **gadgets** inside the flasher facilitate separating and capturing the

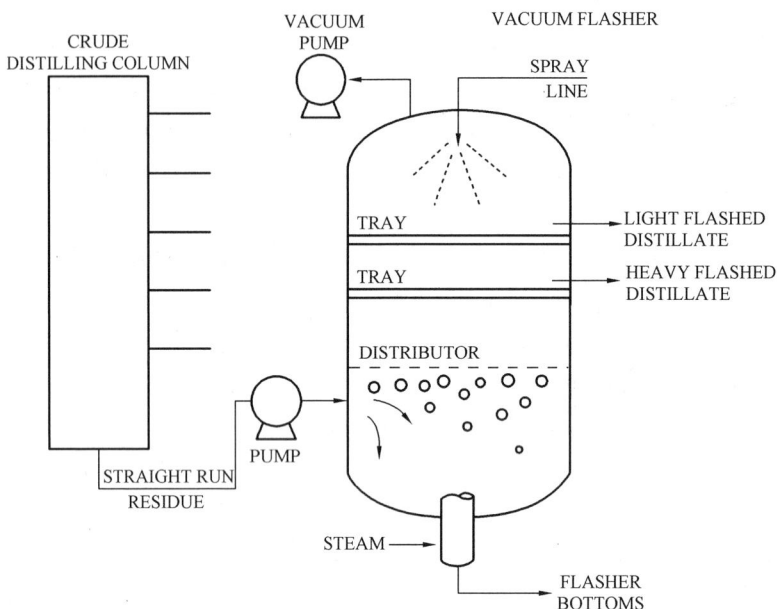

Fig. 3. 9　Vacuum Flashing

offtake streams. As the mixed stream of vapor and liquid enter the flasher, almost all the liquid falls to the bottom. Some of the liquid forms droplets—not quite gaseous but carried along with the vaporized part of the straight run residue. They start to rise with the vapor. To capture these droplets and to disperse the vapor evenly throughout the flasher, the vapor/droplet mixture encounters a distributor—a thick **mesh** screen or a **tray** of loose metal rings several inches thick. The distributor catches the droplets, which drip down to the bottom. Further up the flasher are two trays that operate the same as those in a distilling column. There will probably also be a reboiler, though it's not shown in the figure. **Reflux**, serving the same purpose as it does in a distilling column, is in the form of a liquid spray from the top of the vessel, using some of the cooled streams drawn off one of the trays. The vacuum pump at the top of the vessel maintains the low pressure in the vessel and continuously draws off any vapors that have not **condensed**, and usually consist of small amounts of water and some hydrocarbon. Some vacuum flashers use a device called a steam **eductor** to maintain the low pressure.

The two streams drawn off the upper part of the flasher, light flashed distillate, and heavy flashed distillate, together are called flasher tops. The two streams are often kept **segregated**. Refiners found through chemical analysis that the heavier part of the flasher tops contains most of the **contaminants** that poison **catalysts** in processing units further downstream. By keeping the streams segregated, they can treat the heavy flashed distillate and remove most of the bad actors.

Meanwhile, at the bottom of the flasher, the liquids drain through a pipe. Some flashers have **apparatus** for one last pass at removing the good stuff. They pump some super – heated steam up through the pipe as the liquid comes down, releasing some lighter hydrocarbons that may have been entrained in the liquid. Because the superheated steam is at temperatures well above the cracking temperatures, the liquid coming out of the bottom, the flasher bottoms, is **quenched** with some cooled liquids before the cracking has a chance to take place(cracking requires a minimum residence time to get started).

The flasher bottoms have several destinations—feed to an **asphalt plant**, a thermal cracker, a coker, or as a blending component for residual fuel. The real reason to run a flasher is to make the flasher tops to feed to a cat cracker.

Review

Flashing is the **equivalent** of distilling the straight run residue at a cut point in the neighborhood of $1000 \sim 1100$ ℉. Crude oil distillation curves always show the temperature – volume relationship as if the theoretical distillation took place, i. e. ,they assume the **vacuum flasher** is part of the distillation unit.

Words and Expressions

vaguely	adv. 模糊地
cracking	n. 分解
extensively	adv. 广泛地
specialize in	专门从事
altitude	n. 纬度
illustrate	v. 说明,表明
molecule	n. 分子
transmit	v. 传送
residue	n. 渣油
squat	adj. 矮胖的

gadget	n. 小构件
offtake	adj. 排出的,流出的
mesh	n. 筛网
tray	n. 筛孔
reflux	n. 回流;逆流
condensed	adj. 浓缩的
eductor	n. 喷射器
segregated	adj. 分离的;隔离的
contaminant	n. 杂质
catalyst	n. 催化剂
apparatus	n. 装置
quench	v. 冷却

Technical Terms

vacuum flashing	减压蒸馏
atmospheric pressure	大气压
asphalt plant	沥青装置

Language Focus

1. The discussions so far about distillation curves and distilling columns have treated the shape of the curves at temperatures close to 900 $°F$ somewhat vaguely, but on purpose.

本句中"on purpose"表示有目的的。例如:He knocked the old man down on purpose. 他故意把那个老人撞倒。另外,句中"so far"表示"到现在为止",例如:I have had no reply from her so far. 我至今没有得到她的答复。

2. Since the pressure is lower, water will boil at a lower temperature.

(参考译文为:由于压力的降低,水的沸点也随之下降。)

句中"since"一词表示"原因",不表示"时间"。例如:Since we're all here, we might as well begin our class. 既然大家都来了,我们就上课吧。句中所指的原因通常是提示或补充说明的原因,而不是一定要说的原因。

3. The rate at which they escape depends on the rate at which the heat is delivered (how high you turn the burner) and the pressure of the air above them.

[参考译文为:逸出速率取决于热量传输速率(炉温)和液体上部空气的

压力。]

句中"at which"引导定语从句修饰"the rate",在"which"引导的定语从句中,可以将介宾短语中的介词放到"which"的前面。另外,与"rate"搭配使用的介词为"at",因此"which"前的介词为"at"。

4. The straight run residue leaves the pump and heads down the line and into the flasher, where the pressure has been lowered well below atmospheric pressure.

(参考译文为:直馏渣油离开泵后一路向下进入闪蒸罐,此时其压力已降低到常压以下。)

本句中"where the pressure has been lowered well below atmospheric pressure"引导非限定性定语从句,修饰"flasher"。非限定性定语从句通常由逗号隔开,因其修饰和限定作用并不强,仅起到提示作用。

5. Reflux, serving the same purpose as it does in a distilling column, is in the form of a liquid spray from the top of the vessel, using some of the cooled streams drawn off one of the trays.

本句中"serving the same purpose as it does in a distilling column"为插入语,由现代分词引导,起提示和补充说明的作用,意为"与蒸馏塔的目的一样"。

6. Crude oil distillation curves always show the temperature – volume relationship as if the theoretical distillation took place, i. e. , they assume the vacuum flasher is part of the distillation unit.

(参考译文为:原油的蒸馏曲线总是显示温度和体积的关系,理论上蒸馏似乎是这样的,也就是说,他们认为真空闪蒸塔是蒸馏装置的一部分。)

本句中"as if"表示好像或如同。"as if"后接非事实性描述时用虚拟语气,as if 后接为事实性描述时为正常的陈述语气。例如:

She looks as if she were ten years younger. 她看起来好像年轻了十岁。

It looks as if it is going to rain. 天看起来要下雨。

Reinforced Learning

I . Answer the following questions for a comprehension of the text.

1. What will occur at temperatures close to 900 °F?

2. What is the relationship between pressure and temperature?

3. What is vacuum flashing?

4. What are flasher tops?

5. What is the real reason to run a flasher?

II. Multiple choice: choose the correct one from the alternative answers to give the exact meaning of the word underlined.

1. The discussions <u>so far</u> about distillation curves and distilling columns have treated the shape of the curves at temperatures close to 900 ℉ somewhat vaguely, but on purpose.

    A. far away   B. to some extend

    C. up to now   D. in a distance

2. Suppose you own a chain of restaurants that <u>specialize in</u> serving stuffed baked potatoes—one in Houston, Texas and one in Denver, Colorado.

    A. engage in   B. are special in nature

    C. study   D. cook

3. All this <u>illustrates</u> the simple relationship between pressure and temperatures.

    A. explains   B. shows   C. determines   D. denies

4. The <u>point</u> is, the lower the pressure, the lower the boiling temperature.

    A. reason   B. symbol   C. key   D. problem

5. Various gadgets inside the flasher facilitate separating and <u>capturing</u> the offtake streams.

    A. taking   B. catching   C. keeping   D. finding

6. Some vacuum flashers use a <u>device</u> called a steam eductor to maintain the low pressure.

    A. container   B. tool   C. instrument   D. suggestion

7. <u>Meanwhile</u>, at the bottom of the flasher, the liquids drain through a pipe.

    A. Surprisingly   B. Obviously

    C. Constantly   D. At the same time

8. Flashing is <u>the equivalent of</u> distilling the straight run residue at a cut point in the neighborhood of 1000～1100 ℉.

    A. same to   B. opposite to   C. overlap with   D. similar to

III. Multiple choice: read the four suggested translations and choose the best answer.

1. In the distillation process, all the uncontrolled <u>symptoms</u> of cracking

could occur.

    A. 过程   B. 症状   C. 特点   D. 问题

    2. The lower the pressure, the less energy has to be transmitted, and the lower the temperature at which the vapor will start forming in the liquid, i. e. boil.

    A. 传送   B. 分解   C. 催化   D. 裂化

    3. The two streams are often kept segregated.

    A. 分离   B. 聚合   C. 黏合   D. 吸引

    4. Because the superheated steam is at temperatures well above the cracking temperatures, the liquid coming out of the bottom, the flasher bottoms, is quenched with some cooled liquids before the cracking has a chance to take place.

    A. 排序   B. 沸腾   C. 分解   D. 冷淬

    5. To capture these droplets and to disperse the vapor evenly throughout the flasher, the vapor/droplet mixture encounters a distributor—a thick mesh screen or a tray of loose metal rings several inches thick.

    A. 凝固   B. 加热   C. 散放   D. 冷却

IV. Put the following sentences into Chinese.

    1. The reason is that at the higher altitude of Denver, the atmospheric pressure is lower.

    2. All this illustrates the simple relationship between pressure and temperature. The process of heating lets the molecules absorb enough energy to escape from the liquid form to the gaseous form.

    3. As the mixed stream of vapor and liquid enter the flasher, almost all the liquid falls to the bottom.

    4. Refiners use cracking extensively, but in controlled ways.

    5. Meanwhile, at the bottom of the flasher, the liquids drain through a pipe. Some flashers have apparatus for one last pass at removing the good stuff.

V. Put the following paragraphs into Chinese.

    1. Apply the pressure/boiling relationship to the crude oil cracking problem. The straight run residue will crack if the temperature goes too high, but straight run residue needs to be separated into more cuts. The solution is to do the distillation at reduced pressure.

2. The flasher bottoms have several destinations—feed to an asphalt plant, a thermal cracker, a coker, or as a blending component for residual fuel. The real reason to run a flasher is to make the flasher tops to feed to a cat cracker.

3.3 Recovery of Aromatics and Desulphurisation

⌸ Guidance to Reading

An aromatic hydrocarbon can be monocylic or polycylic. A monocylic aromatic contains one benzene ring with the configuration of six carbon atoms. The term "aromatic" was assigned before the physical mechanism was discovered, and was derived from the fact that many of the compounds have a sweet scent. Sulphur in crude oils can be the simplest compound H_2S, or complex ring structures. The aromatics and sulphur are all harmful to the quality of the mixture they're in, and the aromatics are worth more if they're separated. Therefore, the aromatics should be recovered and sulphur removed.

⌸ Text

Recovery of Aromatics

Removal of **aromatic** compounds can be desirable for two different reasons. Either the aromatics have **detrimental** effects on the quality of the mixture they're in, or the aromatics are worth more if they're separated than if they're not. You have already seen a couple of examples:

- Gasoline now has a max benzene **spec**.
- Aromatic compounds in kerosene cause unacceptable smoke points.

Other more specialized applications include:

- Kerosene range solvents that are aromatics – free or aromatics – laden have various industrial applications.
- Separated benzene, **xylene**, and **toluene** have numerous, chemical applications.
- Removing aromatics from heavy gas oil stocks can improve the lubricating oil characteristics.

Processes

The solvent recovery process is based on the ability of certain compounds to **dissolve** certain classes of other compounds selectively. In this case, certain solvents will dissolve aromatics but not paraffins, olefins, or naphthenes. The reasons the process works are a story you won't read here.

The first **co-requisite** that makes this approach successful is that the solvent with the extracted compounds dissolved in it readily separates itself from the starting hydrocarbon mixture. The second co-requisite is that the solvent and the dissolved extract can easily be split in a **fractionator**.

Take kerosene as an example, one that has a lot of aromatic compounds in it. To half a beaker of kerosene add half a beaker of a solvent—in this case, liquid SO_2. After mixing, the liquid will separate into two phases with the kerosene on the bottom and the SO_2 on top. The kerosene on the bottom will fill less than half the **beaker**. The SO_2, because the aromatic compounds have dissolved in it, will take up more than half the beaker.

If the SO_2 is poured off, the aromatic compounds can be "sprung" from it by simple distillation. This two-step process is batch solvent processing.

Knowing how a simple batch process works, a continuous flow process is easy to **conceptualize**. In Fig. 3. 10, a three column system is shown. The feed is introduced as a vapor into the lower part of a vessel or column with a **labyrinth** of mixers inside (Sometimes the mixers are mechanically moved, such as in a rotating disc contactor). The solvent is introduced as a liquid near the top. The solvent works its way towards the bottom of the vessel, dissolving the extract as it goes along. The rest of the hydrocarbon, which rises to the top, is called **raffinate**.

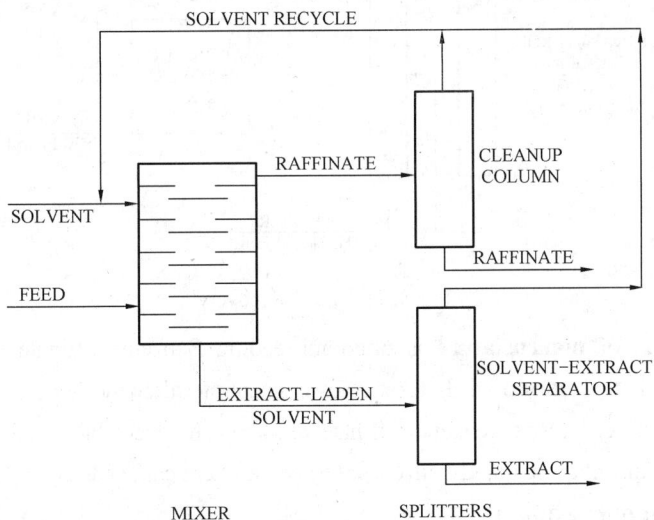

Fig. 3. 10 Solvent Recovery Process

Two columns handle the streams coming out of the mixer. One column cleans up any small amounts of solvent that may have followed along with the raffinate. The other column separates the solvent and the extract. The solvent from both columns is recycled to the top of the mixer.

Some of the solvents used in various applications are listed below:

- Kerosene treating: liquid SO_2, **furfural**.
- Lubricating oil treating: liquid SO_2, Furfural, **phenol**, **propane** (separates **paraffins** from **asphaltenes**).
- Gasoline: **sulfolane**, phenol, **acetonitrile**, liquid SO_2.

Benzene and Aromatics Recovery

The most widespread application of solvent extraction is used in BTX recovery, especially for benzene. To make the process efficient, the feed to the process is pared down to an aromatics concentrate by making a heart cut from a reformate or straight run gasoline stream as shown in Fig. 3. 11. The aromatics concentrate then has a large benzene content, making the extraction process more efficient.

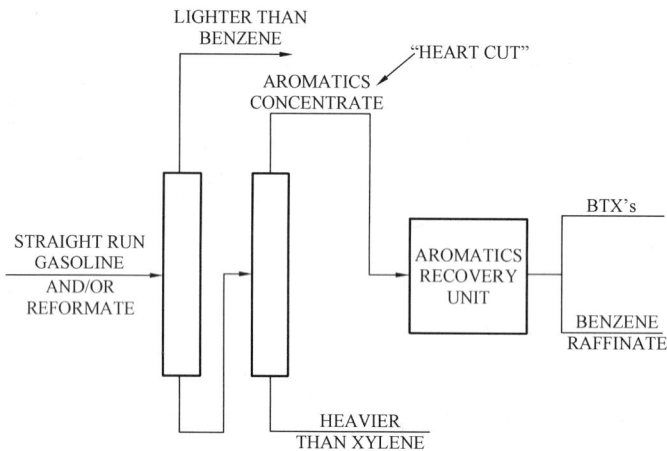

Fig. 3. 11   BTX Recovery

One bit of **nomenclature** is often misleading. Benzene raffinate contains no benzene. It is the leftover of the aromatics concentrate after the goodies (benzene) are removed. As a result, benzene raffinate becomes an acceptable gasoline blending component and the separated benzene becomes a chemical **feedstock**.

Desulphurisation

Sulphur occurs in crude oils combined in a variety of ways, from the simplest compound, H_2S to complex ring structures. H_2S is produced during distillation of the crude oil by decomposition of higher boiling sulphur compounds and appears in the LPG from which it must be removed because of its poisonous and **corrosive** nature. This is done by **counter current washing** with an **amine**, the H_2S being removed for sulphur recovery by heating the amine solution in a separate vessel thus regenerating the amine for recycle to the washing stage. **Mercaptans** can be considered derivatives of H_2S, in which one hydrogen atom is replaced by a carbon/ hydrogen group. Those mercaptans boiling below about 80℃ are readily dissolved in **alkaline solutions** but the **solubility** decreases rapidly above that temperature. For LPG and light gasolines therefore the **mercaptans** can be removed by counter current washing with **caustic soda solution**.

Mercaptans in fractions boiling between 80 and 250℃ cannot be extracted but can be oxidized to disulphides in the Merox solution with air. The disulphides, which are no – corrosive and have little smell, remain dissolved in the oil. Another process for the oxidation of mercaptans uses **copper chloride** as a catalyst. Both processes can be used in the production of aviation jet fuels.

As the cuts taken from crude oil increase in boiling point it is found that the sulphur increases. In the 250 ~ 350℃ range which is used for both diesel fuel and domestic central – heating fuel, the sulphur content is about 1 per cent weight from most Middle East crudes. When this material is burnt the sulphur is oxidized to SO_2 which, being easily oxidized to sulphuric acid, causes atmospheric pollution and corrosion of metals. The sulphur cannot be treated by the methods previously outlines as it is mainly combined with carbon and hydrogen in forms much more complicated than the simple mercaptans. These complex compounds have to be broken down to get at the sulphur, which is done by passing the oil together with hydrogen at high temperature (320 ~ 420℃), over a catalyst containing **cobalt and molybdenum oxides** on **an alumina base**, made in the form of small pellets or extrudates. The reaction is easier and the catalyst life better when the ratio of hydrogen to feed is several times higher than that necessary to complete the reaction chemically. Under these conditions the sulphur compounds decompose and the sulphur combines with the hydrogen to give H_2S. Almost all of the sulphur compounds can be decomposed in this way without affecting the remaining hydrocarbons.

⊡ Words and Expressions

aromatic	adj. 芳烃的
detrimental	adj. 有害的
spec	n. 规格
xylene	n. 二甲苯
toluene	n. 甲苯
dissolve	v. 分解
co – requisite	n. 同时必需的条件
fractionator	n. 分馏器
beaker	n. 烧杯
raffinate	n. 萃余相
conceptualize	v. 形成概念
labyrinth	n. 迷宫
furfural	n. 糠醛
phenol	n. 酚
propane	n. 丙酚
asphaltene	n. 沥青质
paraffin	n. 烷烃
acetonitrile	n. 乙腈
sulfolane	n. 环丁砜
nomenclature	n. 术语
feedstock	n. 原料
corrosive	adj. 腐蚀的
mercaptan	n. 硫醇
amine	n. 胺类
sulphur	n. 硫黄
solubility	n. 可溶性
mercaptan	n. 硫醇
oxidized	adj. 氧化的
disulphide	n. 二硫化物
benzene	n. 苯

⊡ Technical Terms

counter current washing	逆流洗涤

alkaline solution	碱性溶液
copper chloride	氯化铜
caustic soda solution	苛性碱溶液
alumina base	以铝矾土为载体
cobalt oxide	氧化钴
molybdenum oxide	氧化钼

Language Focus

1. Removal of aromatic compounds can be desirable for two different reasons.

本句中"removal"为名词,用名词形式代替动词不定式(to remove)为科技英语中常见的现象。另外,句中"desirable"原意为"有吸引力的",本句中表示"有必要的"。

2. When this material is burnt the sulphur is oxidized to SO_2 which, being easily oxidized to sulphuric acid, causes atmospheric pollution and corrosion of metals.

句中"being easily oxidized to sulphuric acid"为插入语,意为"轻易氧化为硫酸"。另外,句中"which"引导定语从句修饰先行词"sulphuric acid"硫酸。科技英语中插入语较多,意在对概念进行进一步解释说明。

3. The first co – requisite that makes this approach successful is that the solvent with the extracted compounds dissolved in it readily separates itself from the starting hydrocarbon mixture.

句中"solvent with the extracted compounds"为并列关系,意为"溶剂和萃取物"。另外,"is"一词后接"that"表语从句,"that"一般可以省略,但在科技英语中因文体较为正式,"that"多不省略。

4. The SO_2, because the aromatic compounds have dissolved in it, will take up more than half the beaker.

本句的主句为"The SO_2 will take up more than half the beaker"。"take up"意为"占据",表示"二氧化硫占了容器的一半以上"。另外,句中"because the aromatic…"为插入语,对二氧化硫的状态进行充分解释,符合科技语篇的特点。

5. The solvent works its way towards the bottom of the vessel, dissolving the extract as it goes along.

(参考译文为:溶剂在向下流动的过程中发挥作用,溶解与它接触的萃取物。)

句中"work its way"表示溶剂在到达容器底部的过程中一直参与反应。v + 物主代词 + way 表示某个过程中某种行为一直发生,例如:study his way through college 表示一直都在学习的状态。"as it goes along"表示"在溶剂向容器底部流淌的过程中"。

6. To make the process efficient, the feed to the process is pared down to an aromatics concentrate by making a heart cut from a reformate or straight run gasoline stream as shown in Fig. 3. 11.

句中"to"引导的不定式结构表示目的,意为"为了使整个过程更加高效"。另外,句中"pare down"意为削减,例如:We have pared down our expenses to a bare minimum. 我们已最大限度地削减了开支。

7. In this case, certain solvents will dissolve aromatics but not paraffins, olefins, or naphthenes.

句中"in this case"意为"在这种情况下"。例如:Time is an important consideration in this case. 在这种情况下,时间是一个要考虑的重要因素。英语中还有词组 in case,意为"以免",引导让步状语从句,例如:Write the telephone number down in case you forget. 把电话号码写下来,以免忘了。

🔲 Reinforced Learning

I. Answer the following questions for a comprehension of the text.

1. What are the two reasons for the removal of aromatic compounds?

2. What is the solvent recovery process based on?

3. What would happen if the SO_2 is poured off?

4. What should be done to make the solvent extraction efficient?

5. What processes can be used in the production of aviation jet fuels?

II. Multiple choice: choose the correct one from the alternative answers to give the exact meaning of the word underlined.

1. Either the aromatics have detrimental effects on the quality of the mixture they're in, or the aromatics are worth more if they're separated than if they're not.

A. important   B. harmful   C. essential   D. crucial

2. Aromatic compounds in kerosene cause unacceptable smoke points.

A. lead to   B. because   C. avoid   D. keep clear

3. Removing aromatics from heavy gas oil stocks can improve the lubricating oil characteristics.

A. discover   B. increase   C. better   D. destroy

4. If the SO_2 is poured off, the aromatic compounds can be "sprung" from it by simple distillation.

A. see off   B. kick off   C. take off   D. flow out

5. The other column separates the solvent and the extract.

A. gets together   B. sets aparts   C. parallels   D. contains

6. The most widespread application of solvent extraction is used in BTX recovery, especially for benzene.

A. popular   B. advanced   C. fashionable   D. experimental

7. One bit of nomenclature is often misleading.

A. guided   B. directive   C. deceptive   D. provoking

8. Mercaptans in fractions boiling between 80 and 250℃ cannot be extracted but can be oxidized to disulphides (二硫化物) in the Merox solution with air.

A. excluded   B. obtained   C. extend   D. explode

Ⅲ. Multiple choice: read the four suggested translations and choose the best answer.

1. Removal of aromatic compounds can be desirable for two different reasons.

A. 必要的   B. 渴求的   C. 盼望的   D. 有吸引力的

2. To half a beaker of kerosene add half a beaker of a solvent—in this case, liquid SO_2.

A. 大烧杯   B. 溶解器   C. 分解器   D. 加热器

3. The aromatics concentrate then has a large benzene content, making the extraction process more efficient.

A. 提取   B. 浓缩   C. 分解   D. 分裂

4. Mercaptans (硫醇) can be considered derivatives of H_2S, in which one hydrogen atom is replaced by a carbon/ hydrogen group.

A. 副产品   B. 主产品   C. 残余物   D. 衍生品

5. In the 250 ~ 350℃ range which is used for both diesel fuel and domestic central – heating fuel the sulphur content is about 1 per cent weight from most Middle East crudes.

A. 之间   B. 范围   C. 高温   D. 附近

IV. Put the following sentences into Chinese.

1. Kerosene range solvents that are aromatics – free or aromatics – laden have various industrial applications.

2. The first co – requisite that makes this approach successful is that the solvent with the extracted compounds dissolved in it readily separates itself from the starting hydrocarbon mixture.

3. One column cleans up any small amounts of solvent that may have followed along with the raffinate.

4. One bit of nomenclature is often misleading. Benzene raffinate contains no benzene.

5. Sulphur occurs in crude oils combined in a variety of ways, from the simplest compound, H_2S to complex ring structures.

V. Put the following paragraphs into Chinese.

1. The solvent recovery process is based on the ability of certain compounds to dissolve certain classes of other compounds selectively. In this case, certain solvents will dissolve aromatics but not paraffins, olefins, or naphthenes. The reasons the process works are a story you won't read here.

2. Mercaptans in fractions boiling between 80 and 250℃ cannot be extracted but can be oxidized to disulphides (二硫化物) in the Merox solution with air. The disulphides, which are no – corrosive and have little smell, remain dissolved in the oil. Another process for the oxidation of mercaptans uses copper chloride (氯化铜) as a catalyst. Both processes can be used in the production of aviation jet fuels.

Chapter 4   Secondary Refining

4.1   Catalytic Cracking (1)

🔲 Guidance to Reading

To meet the higher and higher demand of gasoline in modern era, one after another cracking techniques have been invented, such as light cracking, thermal cracking, catalytic cracking and so on, among which the most popular one is catalytic. The object of cat cracking process is to convert heavy gas oil to gasoline and lighter. The improvement of catalyst and its reaction section has made the cracking more effective.

🔲 Text

In the **adolescent** years of the petroleum industry, the proportion of the crude oil **barrel** that consumers wanted in the form of gasoline increased faster than fuel oil. It became apparent to refiners that distilling enough crude to make **straight run gasoline** to satisfy the market would result in a **glut** of fuel oils. That would **reverse** the situation in the 19th century when they **literally dumped** gasoline on the ground as they processed enough crude oil to make the fuel oils. The new economy saw increasing prices of gasoline and declining prices of the **heavier cuts**.

To cope with this physical and economic problem, inventive process engineers developed a number of cracking techniques, the most popular of which eventually became cat(catalytic) cracking.

The Process

Here's the process of cat cracking: in a **cat cracker**, straight – run heavy gas oils are subjected to heat and pressure and are contacted with a catalyst to promote cracking.

Definition: A catalyst is a substance added to chemical reaction that **facilitate**s or causes the reaction but when the reaction is complete the catalyst comes out just like it went in. In other words, the catalyst does not change chemically.

It causes reactions between other chemicals.

The cat cracker **comprises the reaction section**, the **regenerator** and the fractionators.

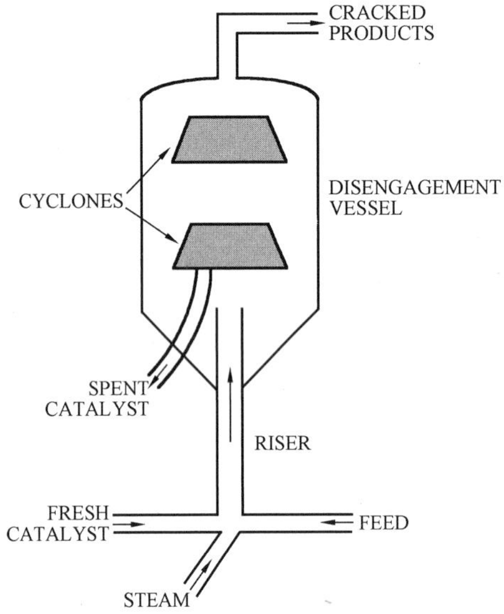

Fig. 4. 1　Cat Cracker Reaction Section

Reaction Section

The **guts** of the cat cracker is the reaction section (Fig. 4. 1), which consists of the cat feed heater and the riser (a pipe from ground level up to a water tank - looking vessel called a **disengagement chamber**). The heater raises the temperature of the cat feed to 900 ~ 1 000 °F ; the feed mixes with catalyst being pumped into the riser. Steam is introduced with the catalyst to give the whole mixture enough lift to climb up to the bottom section of the disengagement chamber. Because of the temperature and the **intimate** contact with the catalyst , all the reactions listed above take place in the riser , even though **residence** time in the riser is only seconds. In the older cat crackers , the disengagement chamber was called the reactor because that's where most of the chemical changes took place. The newer crackers use this vessel only to separate the catalyst from the hydrocarbon.

As the hydrocarbon/catalyst mixture hits the disengagement chamber , it encounters a **cyclone**, a mechanical device that **spins** the mixture. The catalyst being heavier , the **centrifugal** motion slaps it against the walls of the cyclone where it slides to the bottom and out the piping , exiting the disengagement chamber via gravity. The hydrocarbon , mostly in vapor form , but with some liquid droplets , rises from the cyclone to the top of the chamber , encounters cyclone that does the final clean - up act , then exits the top.

Refiners had a purpose in switching to a design where the reactions take place in the riser instead of in the reactor to lengthen the contact time between the feed and the catalyst. However on some riser crackers as the new **vintage** of catalytic cracking units (CCUs) is sometimes called, the hardware is set up to allow some feed to be introduced further up the riser to reduce the contact time. That allows refiners to **segregate** the feed from different crudes, since feeds with different compositions often respond to the catalysts at different rates.

Catalysts

The catalyst used in modern cat crackers is a **marvel** of evolution. It used to be made of natural **alumina – based clay** but now refiners buy only the much— improved **synthetically** produced cat cracking catalysts called **zeolites**. The particles have three unusual characteristics. If you had a jar of cat cracker catalyst and shook or tilted it, the powder, which looks like off – white baby powder, would **slosh** around just like a fluid. This behavior, so important to the design of the whole process, gave rise to another name refiners often use, fluid cat cracking.

The second characteristic is not apparent to the naked eye. Under a microscope you would be able to see that each catalyst particle has a large number of **pores**, and as a consequence, a tremendous surface area, especially in relation to the size of the particle. If the particles were the size of the planet earth, the pores would be deeper than the Grand Canyon and spaced every few miles or so over the surface. The influence of the catalyst depends on contact with the cat feed so the huge surface area is vital to the process.

The third characteristic in the modern cat cracker catalysts comes from technology leaps. The old **alumina – based clay** catalyst had the minerals necessary to promote the cracking reactions. Nowadays the catalysts are synthesized to exacting **dimension** and mineral content. The pores are designed and fabricated so minutely that they will let in just one molecule at a time, and only molecules of a certain size. The way the types of molecules are catalyzed can be controlled. The catalyst suppliers can furnish refiners with catalysts that will favor the creation of high octane gasoline components or, perhaps, the light of olefins. Other catalysts are designed to permit the use of heavier feeds, or not to create so much coke, or to reduce the temperature of the reaction to save energy, and so on.

Words and Expressions

adolescent   adj. 青少年的;青春期的;早期的

barrel	n. 汽油桶
glut	n. 供过于求;过量
reverse	v. 彻底改变;逆转
literally	adv. 按照字面上地;直接地
dump	v. 倾倒
facilitate	v. 促进
comprise	v. 包含
regenerator	n. 再生器
gut	n. 内脏
disengagement	n. 解脱
intimate	adj. 亲密的
residence	n. 居住;停留
cyclone	n. 旋风分离器
spin	v. 旋转
centrifugal	adj. 离心的
vintage	n. 收获
segregate	v. 分离
synthetically	adv. 综合地;合成地
zeolite	n. 沸石;分子筛
slosh	v. 流动
pore	n. 孔
dimension	n. 尺寸

Technical Terms

straight run gasoline	直馏汽油
heavier cuts	重质组块
cat cracker	催化裂化装置
the reaction section	反应器
disengagement chamber	分离器
alumina – based clay	氧化铝基白土(白土催化剂)

Language Focus

1. In the adolescent years of the petroleum industry, the proportion of the crude oil barrel that consumers wanted in the form of gasoline increased faster than fuel oil.

本句中"in the adolescent years of…"中的"adolescent"表示年轻时期，青春期，本课中表示石油工业发展的早期阶段。另"…that consumers wanted in the form of gasoline…"为定语从句修饰先行词"the crude oil"。

2. It became apparent to refiners that distilling enough crude to make straight run gasoline to satisfy the market would result in a glut of fuel oils.

（参考译文为：对炼油厂来说，加工原油生产直馏汽油以满足市场的同时，会造成燃料油的过剩。）

本句为形式主语句，"it"为形式主语，真正的主语是"that distilling enough crude to make straight run gasoline to satisfy the market would result in a glut of fuel oils"。形式主语句用 it 做形式主语，将真正的主语放在后面，起到了平衡句子的作用。

3. The guts of the cat cracker is the reaction section, which consists of the cat feed heater and the riser.

（参考译文为：催化裂化装置的核心部分是反应器，反应器由原料预热装置和提升管反应器组成。）

句中"which consists of the cat feed heater and the riser"为非限定性定语从句，修饰先行词"the reaction section"。另外，句中"consist of"意为"由……组成"，在科技英语中常用来表示组件构成。

4. Because of the temperature and the intimate contact with the catalyst, all the reactions listed above take place in the riser, even though residence time in the riser is only seconds.

本句中"even though"引导让步状语从句，意为"尽管"，例：Even though you do not like it, you must do it. 尽管你不喜欢这工作，你也得做。"residence"原意为"居住"，在本文中为停留，"residence time"指的是物料停留在提升管内反应的时间。

5. The catalyst being heavier, the centrifugal motion slaps it against the walls of the cyclone where it slides to the bottom and out the piping, exiting the disengagement chamber via gravity.

（参考译文为：因为催化剂的密度大，受离心力的作用，催化剂撞击到旋风分离器壁后沿器壁滑落到底部，通过重力的作用脱离沉降器。）

句中"the catalyst being heavier"为独立主格，其作用相当于一个状语从句，表示原因。

6. That allows refiners to segregate the feed from different crudes, since feeds with different compositions often respond to the catalysts at different

rates.

（参考译文为：这是因为成分不同的原料与催化剂的反应速率不同，这种进料方式可以处理不同的原料油。）

句中"since feeds with different compositions often respond to the catalysts at different rates"为原因状语从句。"since"引导的原因状语从句通常是对主句的补充解释。

7. This behavior, so important to the design of the whole process, gave rise to another name refiners often use, fluid cat cracking.

本句中"so important to the design of the whole process"为插入语，对"this behavior"起到补充说明的作用。另外，"give rise to"意为"引发，导致"。例：Her disappearance gave rise to the wildest rumors. 她失踪一事引起了各种流言蜚语。句中"fluid cat cracking"意为"流化催化裂化"。

Reinforced Learning

I. Answer the following questions for a comprehension of the text.

1. What was the problem the refiners came across in the adolescent years of the petroleum industry?

2. What is the process of cat cracking?

3. What does cat cracker comprise?

4. What is the second unusual characteristics of aeolites?

5. What is the difference between the old catalyst and the catalyst used today?

II. Multiple choice: choose the correct one from the alternative answers to give the exact meaning of the word underlined.

1. It became apparent to refiners that distilling enough crude to make straight run gasoline to satisfy the market would result in a glut of fuel oils.

A. urgent   B. obvious   C. challenging   D. critical

2. To cope with this physical and economic problem, inventive process engineers developed a number of cracking techniques, the most popular of which eventually became cat(catalytic)cracking.

A. finally   B. naturally   C. avoidably   D. historically

3. The cat cracker comprises the reaction section, the regenerator and the fractionators.

A. compromises B. composes C. including D. contains

4. A catalyst is a substance added to chemical reaction that <u>facilitates</u> or causes the reaction but when the reaction is complete the catalyst comes out just like it went in.

A. promotes B. hinders C. interrupts D. prevents

5. The guts of the cat cracker is the reaction section(Fig. 4.1), which consists of the cat feed heater and the riser(a pipe from ground level up to a water tank – looking <u>vessel</u> called a disengagement chamber).

A. bottle B. container C. vein D. barrel

6. As the hydrocarbon/catalyst mixture hits the disengagement chamber, it <u>encounters</u> a cyclone a mechanical device that spins the mixture.

A. meets B. conflicts C. confronts D. contacts

7. The catalyst used in modern cat crackers is a <u>marvel</u> of evolution.

A. fantasy B. miracle C. innovation D. illusion

8. Under a microscope you would be able to see that each catalyst particle has a large number of pores, and as a <u>consequence</u>, a tremendous surface area, especially in relation to the size of the particle.

A. matter of fact B. rule

C. result D. whole

9. The influence of the catalyst <u>depends on</u> contact with the cat feed so the huge surface area is vital to the process.

A. relies on B. relates to

C. mixes with D. associates with

10. The third characteristic in the modern cat cracker catalysts comes from technology <u>leaps.</u>

A. developments B. movements C. retreats D. hops

III. Multiple choice：read the four suggested translations and choose the best answer.

1. It became apparent to refiners that distilling enough crude to make straight run gasoline to <u>satisfy the market</u> would result in a glut of fuel oils.

A. 使市场满意 B. 满足市场需求

C. 迎合市场需求 D. 使市场需求满意

2. If the particles were the size of the planet earth, the pores would be deeper than the Grand Canyon and <u>spaced</u> every few miles or so over the surface.

A. 留间隔　　　B. 位于空间　　　C. 留缝隙　　　　D. 忘记

3. If you had a jar of cat cracker catalyst and shook or tilted it, the powder, which looks like off – white <u>baby powder</u>, would slosh around just like a fluid.

A. 婴儿爽身粉　　　　　　B. 婴儿奶粉

C. 婴儿面粉　　　　　　　D. 婴儿蛋白粉

4. That allows refiners to segregate <u>the feed</u> from different crudes, since feeds with different compositions often respond to the catalysts at different rates.

A. 原料　　　B. 填充物　　　C. 食物　　　　D. 喂料

5. Here's the process of cat cracking: in a cat cracker, straight – run heavy gas oils <u>are subjected to</u> heat and pressure and are contacted with a catalyst to promote cracking.

A. 受……制约　　　　　　B. 以……为主题

C. 以……为主语　　　　　D. 屈服于

Ⅳ. Put the following sentences into Chinese.

1. As the hydrocarbon/catalyst mixture hits the disengagement chamber, it encounters a cyclone a mechanical device that spins the mixture.

2. That allows refiners to segregate the feed from different crudes, since feeds with different compositions often respond to the catalysts at different rates.

3. In the older cat crackers, the disengagement chamber was called the reactor because that's where most of the chemical changes took place. The newer crackers use this vessel only to separate the catalyst from the hydrocarbon.

4. However on some riser crackers as the new vintage of catalytic cracking units (CCUs) is sometimes called, the hardware is set up to allow some feed to be introduced further up the riser to reduce the contact time.

5. The influence of the catalyst depends on contact with the cat feed so the huge surface area is vital to the process.

Ⅴ. Put the following paragraphs into Chinese.

1. To cope with this physical and economic problem, inventive process engineers developed a number of cracking techniques, the most popular of which eventually became cat(catalytic) cracking.

2. The third characteristic in the modern cat cracker catalysts comes from

technology leaps. The old alumina – based clay catalyst had the minerals necessary to promote the cracking reactions. Nowadays the catalysts are synthesized to exacting dimension and mineral content. The pores are designed and fabricated so minutely that they will let in just one molecule at a time, and only molecules of a certain size. The way the types of molecules are catalyzed can be controlled.

4.2 Catalytic Cracking (2)

🔲 Guidance to Reading

Some portion of hydrocarbon cracks to coke and ends as a deposit on the catalyst. The spent catalyst must flow to the regenerator to be regenerated. The catalyst need continuous motion going through the cracking/ regeneration cycle. When the cracked product leaves the reaction chamber, it is pumped into a fractionating column dedicated to various products. The cracking process results in the products different in composition than those from the crude distilling column light ends.

🔲 Text

During the cracking process, some portion of hydrocarbon cracks all the way to **coke** and ends as a deposit on the catalyst. As the catalyst surface is covered, the catalyst becomes inactive (spent) and reduces its effectiveness.

To remove the carbon, the spent catalyst flows by gravity to a vessel called a regenerator (Fig. 4.2). Heated air, about 1100 °F, is mixed with the spent catalyst and a chemical reaction takes place:

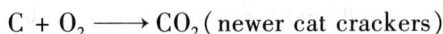

$$C + O_2 \longrightarrow CO \text{ and } CO_2 \text{ (older cat crackers)}$$

$$C + O_2 \longrightarrow CO_2 \text{ (newer cat crackers)}$$

This process, oxidation of coke, is similar to burning coal or **briquettes** in that carbon unites with oxygen and gives off carbon dioxide (CO_2), perhaps carbon **monoxide** (CO), and a large amount of heat. The heat, in the form of the hot CO/CO_2, is generally

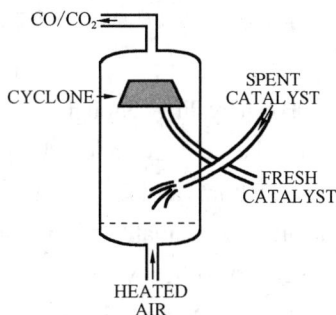

Fig. 4.2 Catalyst Regenerator

used in some other part of the process, such as raising steam to drive pumps or **turbines**. In the older cat crackers, the CO/CO_2 is sent to a CO furnace where oxidation of the rest of tile CO to CO_2 is completed before the CO_2 is finally blown out to the atmosphere.

The regenerator has its own **cyclone** at the top that separates tile catalyst from the CO/CO_2. The regenerated catalyst flows from the cyclone, again by gravity, out of the regenerator, ready to be mixed with cat feed and steam and sent up the riser once more. Thus the catalyst is in continuous motion going through the cracking/ regeneration cycle.

The Fractionator

Meanwhile , back on the hydrocarbon side, when the cracked product leaves the reaction chamber, it is charged(pumped in) to a fractionating column **dedicated** to the cat cracker product (Fig. 4. 3) . The products separated generally are the gases(C_4 and lighter) , cat cracked gasoline, cat cracked light gas oil, cat cracked heavy gas oil, and the fractionator bottoms, called **cycle oil**. A variety of things can be done with the cycle oil but the most popular is to mix it with the fresh cat feed and run it through the reaction again. Some of the cycle oil cracks each time through the reactor.

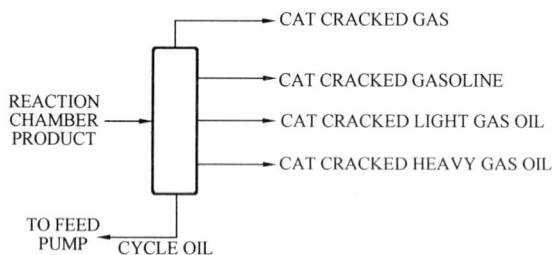

Fig. 4. 3 Fractionation

By recycling enough, all the cycle oil can be made to disappear. The process has the **ominous** designation recycling to **extinction**. Sometimes the most stubborn molecules just keep going around in a circle with no further cracking, so a small amount of cycle oil can be drawn off continuously and blended off to heavy fuel oil.

The cat cracked heavy gas oil can be used as feed to a hydrocracker or thermal cracker or as a residual fuel component. The light gas oil makes a good blending stock for distillate fuel, and the cat cracked gasoline a good motor gasoline blending component.

There is quite a bit of **latitude** in the cut point between the gasoline and light gas oil stream. Refiners use this as one way to regulate the balance between gasoline and distillate as the seasons change. As the winter heating oil season arrives, many refineries go into a max distillate mode. They make adjustments to the CCU fractionator to lower the end point of the cat cracked gasoline to push more volume into the cat light gas oil. In the summer, during a max gasoline mode, the shift is in the other direction.

The products from the fractionator are different in composition than those from the crude distilling column light ends. The cracking process results in the creation of HT, so the C_4 and lighter stream not only contain methane, ethane, propane, and butanes but also hydrogen, **ethylene, propylene, and butylenes**. Because of these extra components, this stream is sent to be separated at the cracked gas plant. This is in contrast to the gas from operations like distilling (and, as discussed later, the hydrotreater, hydrocracker, reformer, and others) where the gases contain only saturated compounds. These end in the sats gas plant for separation. The **isobutane** propylene, and butylenes from the cat cracker will be of special interest when discussing **alkylation**, a process that converts these compounds into gasoline blending components.

The other heavier products also differ in composition. During the cracking process, many of the heavy, complex molecules will crack at the connection between the **aromatics rings** and the side chains. Consequently, the cat cracker products tend to be rich in aromatics, the molecules **replete** with the **benzene ring** somewhere in the structure. That's good news for boosting **octane** numbers and making gasoline, but as you'll find out later, bad news for making jet fuel and diesel fuel.

All pieces fit together as shown in Fig. 4. 4. You can see there are two circular flows going on. On the left side, the catalyst goes through the reaction, is regenerated and gets charged back to the reaction again. On the right side, hydrocarbon comes in and goes out, but the cycle oil provides continuous circulation of at least some of the hydrocarbon components.

Since the cycle oil is recycled to extinction, a simple diagram for cat cracking doesn't even show it as entering or leaving the process. Something more important does show up however, the phenomenon of gain. In the yield structure shown in Table 4. 1, the products coming out add up to 118% of the feed going in. That has nothing to do with the recycle stream but only with the

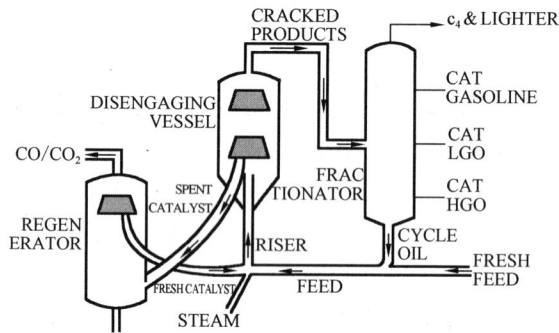

Fig. 4. 4   Cat Cracking Unit

gravities of the products coming out compared to the gravities of the feeds. If you measured the **yields** in percent weight instead of **percent volume**, the yields would come out to 100. But since most US petroleum products are sold by the gallon, US refiners measure everything in volume. Since cracking plays games with the densities, cat cracking yields show a substantial gain. Gain is the **bane** of the accountants but sometimes becomes an **obsession** with refiners as they attempt to "fluff up the barrel" with their cracking processes.

Table 4. 1   Yield Structure

item	% (volume)
Feed: Heavy Gas Oil	40. 0
Flasher Tops	60. 0
Cycle Oil	(10. 0) *
	100
Yield: Coke	8. 0
C_4 and Lighter	35. 0
Cat Cracked Gasoline	55. 0
Cat Cracked Light Gas Oil	12. 0
Cat Cracked Heavy Gas Oil	8. 0
Cycle Oil	(10. 0) *
	118. 0

* Recycle steam not included in feed or yield total.

🔲 Words and Expressions

briquette	n. 坯块;模制试块
monoxide	n. 一氧化物

turbine	n. 涡轮
cyclone	n. 旋风分离器
dedicated	adj. 专注的
ominous	adj. 不好的
extinction	n. 消失
latitude	n. 纬度
olefin	n. 烯烃
ethylene	n. 乙烯
propylene	n. 丙烯
butylene	n. 丁烯
isobutane	n. 异丁烷
alkylation	n. 烷化(作用)
aromatics	n. 芳烃
replete	adj. 充满的;饱食的
octane	n. 辛烷
obsession	n. 困惑
yield	n. 产量
bane	n. 祸患

Technical Terms

benzene ring	苯环
aromatics ring	芳烃环
cycle oil	循环油
percent volume	容积

Language Focus

1. As the catalyst surface is covered, the catalyst becomes inactive (spent) and reduces its effectiveness.

本句中"as"引导状语从句,表示原因,因为催化剂表面被覆盖,导致催化剂活性将下降(待生剂),且催化效率降低。as 引导原因状语从句的情况很多,例如:As the wage of the job was low, there were few applicants for it. 因为工资低,没有什么人申请这份工作。

2. This process, oxidation of coke, is similar to burning coal or briquettes in that carbon unites with oxygen and gives off carbon dioxide(CO_2), perhaps

carbon monoxide(CO), and a large amount of heat.

（参考译文为：积炭的氧化过程与锅炉中煤的燃烧类似，因为碳会和氧结合，产生 CO_2 和大量的热，可能还有一部分 CO 生成。）

句中"oxidation of coke"和"perhaps carbon monoxide(CO)"为插入语，起补充说明的作用。整句的主干为"This process is similar to burning coal or briquettes in that carbon unites with oxygen and gives off carbon dioxide and a large amount of heat"，其中，"in that"表示原因，因为碳会和氧结合，产生二氧化碳和热量。

3. A variety of things can be done with the cycle oil but the most popular is to mix it with the fresh cat feed and run it through the reaction again.

（参考译文为：循环油有多种加工途径，最常见的是与新鲜原料油混合后再次参与裂化反应。）

本句中"a variety of"表示"种种"，意为循环油的处理方式不止于此，但其中最突出的是与新鲜原料油混合后再次参与裂化反应。

4. The light gas oil makes a good blending stock for distillate fuel, and the cat cracked gasoline a good motor gasoline blending component.

本句中"and"为连词，表达平行关系，本句意为"轻柴油是燃料油的优良调和组分，催化裂化汽油是汽油的重要调和组分"，两个小分句是并列的。

5. As the winter heating oil season arrives, many refineries go into a max distillate mode.

本句中"as"表示时间，"随着冬季的来临"。这种用法与之前提到的"as"表示原因的用法应区别对待。

6. That's good news for boosting octane numbers and making gasoline, but as you'll find out later, bad news for making jet fuel and diesel fuel.

本句中"as"意为"正如，就像"。"good news"表示"对……有利"；而"bad news"表示"对……无利"。

🔲 Reinforced Learning

I. Answer the following questions for a comprehension of the text.

1. How to remove the carbon?

2. What are the two circular flows going on in the vessels?

3. What will the refiners do as the winter heating oil season arrives?

4. What do C_4 and lighter stream contain?

5. What is the most popular thing to be done with the cycle oil ?

Ⅱ. Multiple choice：**choose the correct one from the alternative answers to give the exact meaning of the word underlined.**

1. By recycling enough, all the cycle oil can be made to disappear. The process has the ominous designation recycling to extinction.

    A. disappearance   B. explanation

    C. exertion   D. extraversion

2. There is quite a bit of latitude in the cut point between the gasoline and light gas oil stream.

    A. freedom   B. range   C. haphazardry   D. height

3. Refiners use this as one way to regulate the balance between gasoline and distillate as the seasons change.

    A. keep   B. limit   C. restrict   D. discipline

4. The products from the fractionator are different in composition than those from the crude distilling column light ends.

    A. character   B. company   C. element   D. framework

5. Because of these extra components, this stream is sent to be separated at the cracked gas plant.

    A. field   B. company   C. firm   D. factory

6. On the right side, hydrocarbon comes in and goes out, but the cycle oil provide continuous circulation of at least some of the hydrocarbon components.

    A. continual   B. constant   C. irregular   D few

7. If you measured the yields in percent weight instead of percent volume, the yields would come out to 100.

    A. gained   B. explored   C. tested   D. employed

8. Gain is the bane of the accountants but sometimes becomes an obsession with refiners as they attempt to "fluff up the barrel" with their cracking processes.

    A. plump up   B. used up   C. clean up   D. face up

Ⅲ. Multiple choice：**read the four suggested translations and choose the best answer.**

1. Consequently, the cat cracker products tend to be rich in aromatics, the molecules replete with the benzene ring somewhere in the structure.

A. 含有　　　　B. 充满　　C. 完成　　　　D. 重塑

2. This is in contrast to the gas from operations like distilling(and, as discussed later, the hydrotreater, hydrocracker, reformer, and others) where the gases contain only <u>saturated</u> compounds.

A. 饱和的　　　B. 不饱和的　C. 半饱和的　　D. 无机的

3. Meanwhile, back on the hydrocarbon side, when the cracked product leaves the reaction chamber, it is charged(pumped in)to a fractionating column <u>dedicated to</u> the cat cracker product.

A. 用于　　　　B. 致力于　　C. 奉献给　　　D. 投身于

4. In the older cat crackers, the CO/CO_2 is sent to a CO furnace where oxidation of the <u>rest</u> of tile CO to CO_2 is completed before the CO_2 is finally blown out to the atmosphere.

A. 剩余部分　　B. 休眠状态　C. 不活跃部分　　D. 活跃部分

5. Something more important does <u>show up</u> however, the phenomenon of gain.

A. 展示　　　　B. 呈现出来　C. 炫耀　　　　D. 展览

Ⅳ. Put the following sentences into Chinese.

1. The heat, in the form of the hot CO/CO_2, is generally used in some other part of the process, such as raising steam to drive pumps or turbines.

2. Thus the catalyst is in continuous motion going through the cracking/ regeneration cycle.

3. Meanwhile, back on the hydrocarbon side, when the cracked product leaves the reaction chamber, it is charged(pumped in)to a fractionating column dedicated to the cat cracker product.

4. There is quite a bit of latitude in the cut point between the gasoline and light gas oil stream.

5. The products from the fractionator are different in composition than those from the crude distilling column light ends.

Ⅴ. Put the following paragraphs into Chinese.

1. The cat cracked heavy gas oil can be used as feed to a hydrocracker or thermal cracker or as a residual fuel component. The light gas oil makes a good blending stock for distillate fuel, and the cat cracked gasoline a good motor gasoline blending component.

2. All pieces fit together as shown. You can see there are two circular flows going on. On the left side, the catalyst goes through the reaction, is regenerated and gets charged back to the reaction again. On the right side, hydrocarbon comes in and goes out, but the cycle oil provide continuous circulation of at least some of the hydrocarbon components.

4.3 Catalytic Reforming

Guidance to Reading

Catalytic reforming is an important process for upgrading low octane naphtha to a high octane gasoline blending component, reformate, either increasing the quality or the quantity of the gasoline. But the benzene content of gasoline that came primarily from the cat reformers became the focus of environmental concern, giving rise to the debates between refiners and environmental protectors.

Text

Cat reforming, has provided more **controversy** in the refining business than all the other units **combined**. What started out as an engineering solution to market needs ended in the middle of debates between refiners and environmental protectors.

The History

File tortuous – history began when refiners created cat reformers as a method of raising both the volume and quality of gasoline. In the first half of the 20th century the demand for gasoline grew at twice the rate of fuel oils. The **specifications** of the gasoline that car manufacturers designed into their vehicles increased incessantly. America built four and six lane highways across the continent and consumers moved beyond Henry Ford's Model T to big, fat comfortable passenger cars. Their engines needed cheap higher **octane** gasoline.

As the refiners scrounged around their refineries looking for suitable blending components, they found heavy naphtha that was purposely left in kerosene. Catalytic reforming could upgrade the quality of these naphthas, some with octane numbers in the range of 35 to 40, to as much as 90 octane, increasing the quality and quantity of the gasoline—making capacity at the same time.

The Golden Years. In 1949, Universal Oil Products Company introduced the present day catalyst and plant design. They arrived just in time for the Golden Years, so – called because of the quarter Century following World War II of HT 7% economic growth in the US throughout this period, refiners found themselves in an octane race, competing by advertising gasoline with ever – higher octane numbers. The gas pump with 100 octane **premium** gasoline seemed hydrocarbon's Holy Grail. As a way to manage the truth in advertising, more refines built new and improved cat reformers, boosting the quality of their gasoline even further.

The environment. In the 1970s, enough concern about the environment reached the public **agenda** that governments began requiring refiners to reduce the amount of lead put in gasoline. For decades, refiners had exploited the mysterious fact that the addition of tiny amounts of **tetraethyl** lead could substantially increase the **gasoline octane**. They had always known that lead was toxic, but in the mid – 1970s the government published a schedule to phase out in a period of 10 years the already restricted amount of lead allowed. Refiners built new cat reformers, **debottlenecked** old ones, and introduced new catalysts into their existing ones to increase the amount of aromatics, which were high octane gasoline blending components.

Beddour's Law—the premise that one cannot eliminate a pollutant without creating another one—came into effect as lead phased out in the 1980s. The benzene content of gasoline became a big issue. By then benzene was well known as a **carcinogen** and even the few percent that came primarily from the cat reformers became the focus of concern and eventually phase – down regulation. Once again the refiners went to work to change the feeds and modify the catalysts in the reformer to eliminate this pollutant, replacing it with some other benign, **albeit mutated** (cat reformed) molecules. Complicating the effort were the limits put on the total aromatics content, since these compounds, with the characteristic benzene ring imbedded in them, were suspect as well.

The **by – product** workhorse. Finally, throughout the evolution of cat reforming, refiners became increasingly aware that one of the reformer by – products, hydrogen, emerged as an essential workhorse in the refinery. Hydrogen was increasingly used in hydrotreaters to help remove sulfur and other **contaminants** from various dirty streams and change the structure of some others. Even more, hydrocrackers had become a black hole for hydrogen. Now when the cat reformer shuts down for maintenance or has an emergency shutdown, major parts of the refinery get affected by the shortage of hydrogen and have to shut down as well.

The good reactions that take place in the cat reforming process are mainly:

- **Paraffins** are converted to **isoparaffins**.
- Paraffins are converted to **naphthenes**, releasing hydrogen.
- Naphthenes are converted to aromatics, releasing hydrogen.

Some not – so – good reactions, from an octane point of view, take place:

- Some of the paraffins and naphthenes crack and form butanes, and lighter gases.
- Some of the side chains get broken off the naphthenes and aromatics and form butanes and lighter gases.

The important thing for you to remember is paraffins and naphthenes get converted to aromatics and **isomers**.

The Hardware

You might expect some unusual hardware would be required to cause these complicated reactions to take place. On the contrary, what's needed is an unusual catalyst, and in this case it's made of alumina, **silica**, **platinum**, and sometimes **palladium**. The platinum is in no small amounts (several million dollars worth in one process unit), so great care is taken to keep track of it. The platinum and palladium are the key ingredients that do the wonderful job of causing paraffins to wrap themselves around in circles and to lose their extra hydrogens and to poke side chains out where none existed before.

There are several ways of putting the hydrocarbon in contact with the catalyst. The one covered here is called fixed bed because the hydrocarbon is dribbled through the catalyst, which stays put in a vessel, or actually several vessels used in a series, as shown in Fig. 4. 5.

The naphtha feed is **pressurized**, heated, and charged to the first reactor, where it **trickles** through the catalyst and out the bottom of the reactor. This process repeats in the next two reactors. The product then runs through a cooler where much of it is liquefied. The purpose of the liquefaction at this point is to permit separation of the hydrogen rich gas stream for recycling. This process is

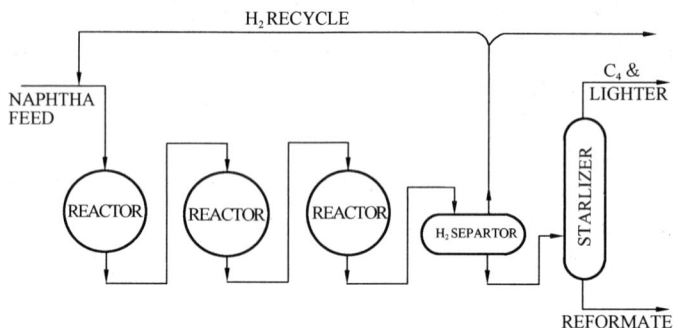

Fig. 4. 5   Catalytic Reforming

important enough to warrant a few sentences.

Meanwhile outside the reactors, part of the hydrogen stream is recycled to the feed while the other part is sent to the gas plant. The liquid product from the bottom of the separator is sent to a fractionator called a **stabilizer**, a column nothing more than a **debutanizer**. It makes a bottom product called reformate; butanes and lighter go overhead and head towards the gas plant.

Review

Catalytic reforming is an important process for upgrading **low octane naphtha** to a high octane gasoline blending component, reformate. The paraffins are converted to isoparaffins and naphthenes are converted to aromatics to capitalize on their higher octane numbers. Unfortunately, the higher the octane number of the reformate, the lower the yield and the more light ends produced.

🔲 **Words and Expressions**

controversy	n. 争议
combined	adj. 组合的
specification	n. 规格,详述
octane	n. 辛烷
sustain	v. 维持
premium	n. 值
agenda	n. 日程
tetraethyl	n. 四乙基
debottleneck	v. 排除故障
carcinogen	n. 致癌物
albeit	conj. 尽管

mutate	v. 突变
contaminant	n. 杂质
paraffin	n. 石蜡
isoparaffin	n. 异链烷烃
naphthene	n. 环烷属烃
isomer	n. 同分异构体
silica	n. 硅石
platinum	n. 铂
palladium	n. 钯
pressurize	v. 加压
trickle	n. /v. 滴流
debutanizer	n. 脱丁烷塔
stabilizer	n. 稳定器

Technical Terms

by – product	副产品
file tortuous – history	曲折历史
low octane naphtha	低辛烷值的石脑油
gasoline octane	汽油辛烷值

Language Focus

1. What started out as an engineering solution to market needs ended in the middle of debates between refiners and environmental protectors.

(参考译文为:起初是以解决市场需求为目的,最终演化成了一场炼油工作者和环境保护者之间的辩论。)

本句中"What started out as an engineering solution to market needs"引导名词性从句在句中作主语。

2. In the first half of the 20th century the demand for gasoline grew at twice the rate of fuel oils.

本句中"grew at twice the rate of fuel oils"意为"比燃料油快两倍的速率增长"英语中的倍数表达法之一:倍数 + the size(length,height. . .)of.

This street is four times the length of that street. 这条街是那条街的 4 倍长。

The height of this hill is four times that of that small one. 这座山的高度是那座小山的 4 倍。

3. As the refiners scrounged around their refineries looking for suitable blending components, they found heavy naphtha that was purposely left in kerosene.

（参考译文为：在寻找合适的组分作为原料的过程中，人们发现从煤油中分离出的重石脑油最合适。）

句中"as"引导时间状语从句，表示主句的动作发生在从句的动作过程中。

4. As a way to manage the truth in advertising, more refines built new and improved cat reformers, boosting the quality of their gasoline even further.

本句中"As a way to manage the truth in advertising"意为作为实现广告承诺的一种方式。"as"在本句中意为"作为"的这一用法应该与前面提到的用法有所区分。

5. Refiners built new cat reformers, debottlenecked old ones, and introduced new catalysts into their existing ones to increase the amount of aromatics, which were high octane gasoline blending components.

（参考译文为：炼油工程师重新设计了催化重整工艺，消除了旧工艺中不足之处，同时开发了一种新的催化剂用于现行催化重整工艺，来增加汽油中芳烃的含量，这是高辛烷值汽油中重要的调和组分。）

句中"which"引导非限定性定语从句，修饰"aromatics"。

6. Now when the cat reformer shuts down for maintenance or has an emergency shutdown, major parts of the refinery get affected by the shortage of hydrogen and have to shut down as well.

本句中"as well"表示"也"的意思。应注意区分"as well as"，"as well"和"might as well"，它们分别表示"和"，"也"以及"还是……的好"。

7. The platinum and palladium are the key ingredients that do the wonderful job of causing paraffins to wrap themselves around in circles and to lose their extra hydrogens and to poke side chains out where none existed before.

（参考译文为：金属铂和钯在催化剂中是最重要的组分，能够促使正构烷烃卷曲成环，脱去多余的氢，拨开以前没有的侧链。）

句中"that do the wonderful job of causing..."为定语从句，修饰"ingredients"。"do the wonderful job"表示这些主要成分所起的重要作用。

8. Meanwhile outside the reactors, part of the hydrogen stream is recycled to the feed while the other part is sent to the gas plant.

本句中"while"表示对比，意为"同时在反应器的外面，一部分氢气重新循环进原料里,其他部分则输送到气体处理装置"。

Reinforced Learning

I. Answer the following questions for a comprehension of the text.

1. What has cat reforming caused in the refining business?

2. What did Universal Oil Products Company introduce in 1949?

3. What did the government ask the refiners to do to protect the environment?

4. What has become a big problem in 1980s?

5. What is an essential workhorse in the refinery? What are its functions?

II. Multiple choice: choose the correct one from the alternative answers to give the exact meaning of the word underlined.

1. Cat reforming, has provided more controversy in the refining business than all the other units combined.

    A. discussion   B. conversation

    C. conservative   D. counterfeit

2. The specifications of the gasoline that car manufacturers designed into their vehicles increased incessantly.

    A. informations   B. requirements

    C. standards   D. advantages

3. As the refiners scrounged around their refineries looking for suitable blending components, they found heavy naphtha that was purposely left in kerosene.

    A. gradually   B. unexpectedly

    C. randomly   D. on purpose

4. The Golden Years. In 1949, Universal Oil Products Company introduced the present day catalyst and plant design.

    A. invented   B. developed   C. planned   D. advertised

5. In the 1970s, enough concern about the environment reached the public agenda that governments began requiring refiners to reduce the amount of lead put in gasoline.

    A. attention to   B. worry about

    C. hesitation on   D. anxiety about

6. By then benzene was well known as a carcinogen and even the few percent that came primarily from the cat reformers became the focus of concern and eventually phase – down regulation.

    A. accidentally   B. finally   C. hopefully   D. obviously

7. Finally, throughout the evolution of cat reforming, refiners became increasingly aware that one of the reformer by – products, hydrogen, emerged as an essential workhorse in the refinery.

    A. necessary   B. alternative   C. basic   D. important

8. Now when the cat reformer shuts down for maintenance or has an emergency shutdown, major parts of the refinery get affected by the shortage of hydrogen and have to shut down as well.

    A. closes down   B. turns off

    C. breaks down   D. puts down

9. On the contrary, what's needed is an unusual catalyst, and in this case it's made of alumina, silica, platinum, and sometimes palladium.

    A. in this regard   B. in this sense

    C. in this way   D. in this section

10. The product then runs through a cooler where much of it is liquefied.

    A. in a liquid state   B. in a gas state

    C. in a solid state   D. in a frozen state

Ⅲ. Multiple choice：read the four suggested translations and choose the best answer.

1. What started out as an engineering solution to market needs ended in the middle of debates between refiners and environmental protectors.

    A. 争论   B. 争吵   C. 争斗   D. 争执

2. As a way to manage the truth in advertising, more refines built new and improved cat reformers, boosting the quality of their gasoline even further.

    A. 提高   B. 鼓吹   C. 宣传   D. 增加

3. Once again the refiners went to work to change the feeds and modify the catalysts in the reformer to eliminate this pollutant, replacing it with some other benign, albeit mutated (cat reformed) molecules.

    A. 取缔   B. 消除   C. 禁止   D. 减少

4. There are several ways of putting the hydrocarbon in contact with the

catalyst.

A. 使……添加   B. 使……分离

C. 使……接触   D. 使……沟通

5. Even more, hydrocrackers had become a black hole for hydrogen.

A. 更有甚者   B. 更多的是

C. 更意外的是   D. 更快的是

IV. Put the following sentences into Chinese.

1. America built four and six lane highways across the continent and consumers moved beyond Henry Ford's Model T to big, fat comfortable passenger cars.

2. As a way to manage the truth in advertising, more refines built new and improved cat reformers, boosting the quality of their gasoline even further.

3. For decades, refiners had exploited the mysterious fact that the addition of tiny amounts of tetraethyl lead could substantially increase the gasoline octane.

4. The product then runs through a cooler where much of it is liquefied. The purpose of the liquefaction at this point is to permit separation of the hydrogen rich gas stream for recycling.

5. You might expect some unusual hardware would be required to cause these complicated reactions to take place.

V. Put the following paragraphs into Chinese.

1. Meanwhile outside the reactors, part of the hydrogen stream is recycled to the feed while the other part is sent to the gas plant. The liquid product from the bottom of the separator is sent to a fractionator called a stabilizer, a column nothing more than a debutanizer. It makes a bottom product called reformate; butanes and lighter go overhead and head towards the sats gas plant.

2. Catalytic reforming is an important process for upgrading low octane naphtha to a high octane gasoline blending component, reformate. The paraffins are converted to isoparaffins and naphthenes are converted to aromatics to capitalize on their higher octane numbers. Unfortunately, the higher the octane number of the reformate, the lower the yield and the more light ends produced.

4. 4 Hydrocracking

🔲 Guidance to Reading

Hydrocracking was designed to accomplish more of what each of the earlier processes (thermal or cat cracking or cat reforming) do. Moreover, hydrocrackers produce no bottom – of – the – barrel leftovers, with the outturn being all light oils. Hydrocrackers can swing refinery yields by maximizing diesel and distillate fuels in the winter and maximizing gasoline and maybe even jet fuel in the summer, meanwhile improving the quality. In addition, the hydrocracker feeds the alky plant with isobutane and the cat reformer with naphtha and gets hydrogen back from the cat reformer.

🔲 Text

Hydrocracking is a process of more recent **vintage** than thermal or cat cracking or cat reforming, but it was designed to accomplish more of what each of the earlier processes do. Plopped in the middle of a refinery, hydrocracking, can take care of many refiners headaches that happen as the market changes from month – to – month or season – to – season. Hydrocrackers can produce gasoline components from light or heavy gas oils. Their quality is better than if these gas oils were recycled through the cracking process that generated them. Hydrocrackers can produce light distillates (jet fuel and diesel fuel) from heavy gas oils. Hydrocrackers produce a relatively large amount of isobutane, a handy supply for the **alky plant**. Maybe best of all, hydrocrackers produce no bottom – of – the – barrel leftovers (coke, pitch, or resid). The outturn is all light oils. Refiners use hydrocrackers to move from max diesel and distillate fuels in the winter to max gasoline and maybe even jet fuel in the summer. In the refineries where they reside, hydrocrackers have become refiners' swing units.

The Process

Hydrocracking is simple. It's catalytic cracking in the presence of hydrogen. Various combinations of hydrogen, catalyst, and operating conditions permit cracking a wide range of feedstocks, from light gas oil to straight run residue or the cycle oils from the cat cracker or thermal cracker. The hydrocracker is run in stages, with each one upgrading to the next cut—heavy stuff to middle distillates, middle distillates to gasoline range components.

Hydrocracking can simultaneously improve the quality of both the gasoline blending and the distillate fuel blending pools. The worst of the distillate stocks, the cracked gas oils, have high aromatics contents that give them poor diesel fuel performance. Passing them through a hydrocracker results in gasoline components with relatively high octane numbers or naphthas that make excellent cat reformer feed.

Why doesn't every refinery have one of these machines? Even though about a dozen different types of hydrocrackers are presently popular, they all are expensive to build and operate. The units described below are typical of most of them.

The Hardware and the Reactions

Hydrocracking catalysts are fortunately less precious and expensive than reforming catalysts. Usually they are compounds of sulfur with **cobalt**, **molybdenum**, or nickel plus alumina (You may have wondered what anyone used those metals for). In contrast to cat cracking, but like cat reforming, hydrocrackers have their catalysts in a fixed bed. Like cat reforming, the process is carried out in more than one reactor—two in the illustration in Fig. 4.6.

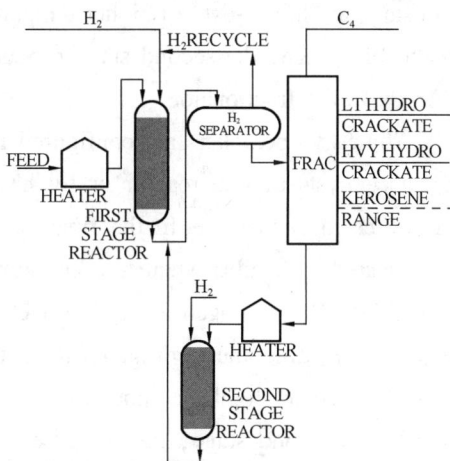

Fig. 4.6 Two Stage Hydrocracker

Feed, in this case cracked heavy gas oil, is mixed with hydrogen vapor, heated to 550 ~ 750 °F, pressurized to 1,200 ~ 2,000 psi, and charged to the first stage reactor. As it passes through the catalyst, about 40% ~ 50% of the feed is cracked to gasoline range material(below 400 °F end point).

The hydrogen and the catalyst are **complementary** in several ways. Tile catalyst causes long chain molecules to crack and the rings in aromatic compounds to open. Both these reactions need heat to keep them going. They are both an **endothermic** process. On the other hand, as the cracking takes place the excess hydrogen floating around saturates the molecules, a process that gives off heat. This process, called hydrogenation, is **exothermic**. Thus, hydrogenation gives off the heat necessary to keep the cracking and ring opening going.

After the hydrocarbon leaves the first stage, it is cooled and liquefied and run through a hydrogen separator. In the **hydrogen separator** the hydrogen is split out for recycle to the feed. The liquid is charged to a fractionator. Whatever products are desired (gasoline components, jet fuel, and gas oil), the fractionator cuts them out of the first stage reactor effluent, leaving a bottom stream ready for the next, second stage, reactor. In other words, kerosene range and light gas oil range material could be taken as separate side draw products from the fractionator, or could be included in the fractionator bottoms if further conversion to gasoline range material is the object.

The bottom stream is again mixed with a hydrogen stream and charged to the second stage. Since this material has already been subjected to some hydrogenation, cracking, and reforming in the first stage, the operations of the second stage are more severe (higher temperatures and pressures). Like the outturn of the first stage, the second stage product is separated from the hydrogen and charged to the fractionator.

Some hydrocrackers are configured in three stages, either by stacking different catalysts in one reactor or by having three reactors. Different reactions take place in each stage. In the first stage, the catalyst opens up the rings of complicated molecules where the contaminates, sulfur and nitrogen, might be **embedded**. The hydrogen forms hydrogen **sulfide** (H_2S) and **ammonia** (NH_3). At the same time the hydrogen fills out many of the opened double bonds, forming simpler, lighter compounds.

In the second stage, run more severely, the catalyst opens more rings of the heavy, complicated molecules and cracks others, forming light products. In the third stage, a kind of polishing job takes place where the olefins and aromatic compounds are saturated, forming naphthene, paraffins, and especially isoparaffins.

Hydrocracker reactors operate at really severe conditions—2,000 psi and 750 °F. Imagine the hardware necessary to contain the reactions. The specialty – steel reactor walls are sometimes 6 inches thick. A critical worry is the possibility of runaway cracking. Since the overall process is **endothermic**, the temperature can rise rapidly, accelerating the cracking rates dangerously, which only gives off more heat as hydrogenation then takes place. Elaborate **quench systems** are built into most hydrocrackers to control runaway.

Residue Hydrocracking

There are a few hydrocrackers that have been constructed to handle straight run residue or flasher bottoms as feed. Most of them are operated as **hydrotreaters**. The yields are over 90% residue fuel. The purpose of the operation is to remove sulfur by the catalytic reaction of the hydrogen and the sulfur compounds, forming H_2S hydrogen sulfide. Residue with a sulfur content of about 4% or less can be converted to heavy fuel oil with less than 0.3% sulfur.

Review

With the addition of the hydrocracker to the refinery processing **scheme**, the absolute requirement for integrated operations becomes apparent. The hydrocracker is a pivot point since it can swing refinery yields between gasoline, distillate fuel, and jet fuel and simultaneously improve the quality. How the hydrocracker runs depends intimately on the feed rates and operating conditions of the cat cracker, the coker or thermal cracker. In addition, the hydrocracker feeds the alky plant with **isobutane** and the cat reformer with naphtha and gets hydrogen back from the cat reformer.

Words and Expressions

vintage	n. 工艺
cobalt	n. 钴
molybdenum	n. 钼
complementary	adj. 补充的;互补的
endothermic	adj. 吸热的
exothermic	adj. 发热的
hydrotreater	n. 加氢处理装置
embed	v. 嵌入
sulfide	n. 硫化物
ammonia	n. 氨
scheme	n. 计划
isobutane	n. 异丁烷

Technical Terms

alky plant	烷化工厂
quench system	急冷系统
hydrogen separator	氢分离器

□ Language Focus

1. Since the overall process is endothermic, the temperature can rise rapidly, accelerating the cracking rates dangerously, which only gives off more heat as hydrogenation then takes place.

（参考译文为：因为所有的过程都是吸热反应，快速升温可以促进裂化反应的进行，只有发生加氢反应后才能放出更多的热。）

在句中"accelerating the cracking rates dangerously"为伴随状语，意为"加速裂化过程"。另外，"which only gives off more heat as hydrogenation then takes place"为"which"引导的非限定性定语从句，修饰之前提到的整个事件。

2. Hydrocrackers produce a relatively large amount of isobutane, a handy supply for the alky plant.

（参考译文为：加氢裂化还可以生产相对大量的异丁烷，可以为烷基化装置提供原料。）

本句中"handy"意为"方便的"，"现成的"，用来形容供应的便捷。诸如此类的用法还有：Our flat is very handy for the schools. 我们的住所离学校很近，非常方便。

3. Various combinations of hydrogen, catalyst, and operating conditions permit cracking a wide range of feedstocks, from light gas oil to straight run residue or the cycle oils from the cat cracker or thermal cracker.

（参考译文为：通过氢气、催化剂和适当操作条件可以加工各种原料，如轻瓦斯油、直馏渣油，催化裂化或热裂化循环油。）

句中"range"表示范围，通常和"from"连用，表示"范围涉及……"。例如：The children's ages range from 8 to 15. 这些孩子们的年龄在 8 岁到 15 岁之间。

4. Even though about a dozen different types of hydrocrackers are presently popular, they all are expensive to build and operate. The units described below are typical of most of them.

（参考译文为：尽管目前大约有 12 种加氢裂化工艺很受欢迎，但是这些装置的建造和操作成本非常昂贵。）

句中"even though"引导让步状语从句,译为"尽管"。例如:Even though I hadn't seen my classmate for many years, I recognized him immediately. 即使多年没有看见我的同学,我也马上认出了他。

5. The liquid is charged to fractionators.

本句中"be charged to"意为"派送"。例如:A specialist has been charged to see to this business. 一个专家已被派来负责此事。

6 In the first stage, the catalyst opens up the rings of complicated molecules where the contaminates, sulfur and nitrogen, might be embedded.

(参考译文为:在第一阶段,催化剂把含有硫、氮和污染物等交织在一起的复杂分子的环打开。)

句中"in the first state"表示"在第一阶段"。英语中表示阶段的词有"section","phase"以及"step"等。另外,"open up"意为"打开"。介词"up"表示动作发生以后的结果,如"eat up, used up, clean up"等。

7. Hydrocracking is a process of more recent vintage than thermal or cat cracking or cat reforming, but it was designed to accomplish more of what each of the earlier processes do.

句中"recent"译为"近期的",表示该工艺是较新的,前沿的工艺。另外,"what each of the earlier processes do"为"what"引导的名词性从句,作"of"的宾语。

Reinforced Learning

I. Answer the following questions for a comprehension of the text.

1. What is hydrocracking?

2. What do refiners use hydrocrackers to do in winter as well as in summer?

3. What can hydrocracking improve immediately during the course?

4. How are the hydrogen and the catalyst complementary to each other?

5. What do the refineries worry most about the hydrocracker reactors?

II. Multiple choice:choose the correct one from the alternative answers to give the exact meaning of the word underlined.

1. It was designed to underline{accomplish} more of what each of the earlier processes do.

A. accompany   B. complete   C. polish   D. eliminate

2. Their quality is better than if these gas oils were recycled through the cracking process that <u>generated</u> them.

    A. abandoned   B. utilized   C. exploited   D. produced

3. With the addition of the hydrocracker to the refinery processing scheme, the absolute requirement for integrated operations becomes <u>apparent</u>.

    A. obvious   B. temporary

    C. ever – lasting   D. often – changing

4. The <u>outturn</u> is all light oils.

    A. turnover   B. outtake   C. output   D. turnup

5. Since this material has already <u>been subjected</u> to some hydrogenation, cracking, and reforming in the first stage, the operations of the second stage are more severe (higher temperatures and pressures).

    A. received   B. removed   C. involved   D. decomposed

6. Some hydrocrackers are <u>configured</u> in three stages, either by stacking different catalysts in one reactor or by having three reactors.

    A. arranged   B. figured   C. composed   D. confined

7. Hydrocracker reactors operate at really <u>severe</u> conditions – 2,000 psi and 750 °F.

    A. plain   B. serious   C. casual   D. impossible

8. A <u>critical</u> worry is the possibility of runaway cracking.

    A. crucial   B. elementary   C. useless   D. hopeless

9. Since the overall process is <u>endothermic</u>, the temperature can rise rapidly, accelerating the cracking rates dangerously, which only gives off more heat as hydrogenation then takes place.

    A. complicated   B. heat – absorbing

    C. often – changing   D. unpredictable

10. <u>Elaborate</u> quench systems are built into most hydrocrackers to control runaway.

    A. complex   B. beautiful   C. delicate   D. strong

III. Multiple choice: read the four suggested translations and choose the best answer.

1. How the hydrocracker runs depends <u>intimately</u> on the feed rates and operating conditions of the cat cracker and residue reduction units the coker or thermal cracker.

A. 密切地　　　　B. 部分地　　　　C. 相对地　　　D. 绝对地

2. The hydrogen and the catalyst are <u>complementary</u> in several ways.

A. 互补的　　　　B. 排斥的　　　　C. 无关的　　　D. 相关的

3. The hydrocracker is a <u>pivot</u> point since it can swing refinery yields between gasoline, distillate fuel, and jet fuel and simultaneously improve the quality.

A. 关键的　　　　B. 相关的　　　　C. 随机的　　　D. 必要的

4. They are both an endothermic process. the temperature can rise rapidly, <u>accelerating</u> the cracking rates dangerously, which only gives off more heat as hydrogenation then takes place.

A. 加速　　　　　B. 减缓　　　　　C. 关联　　　　D. 影响

5. Hydrocracking can <u>simultaneously</u> improve the quality of both the gasoline blending and the distillate fuel blending pools.

A. 帮助　　　　　B. 快速　　　　　C. 有效　　　　D. 同时

Ⅳ. Put the following sentences into Chinese.

1. Their quality is better than if these gas oils were recycled through the cracking process that generated them.

2. Various combinations of hydrogen, catalyst, and operating conditions permit cracking a wide range of feedstocks, from light gas oil to straight run residue or the cycle oils from the cat cracker or thermal cracker.

3. Even though about a dozen different types of hydrocrackers are presently popular, they all are expensive to build and operate.

4. Both these reactions need heat to keep them going. They are both an endothermic process.

5. The bottoms stream is again mixed with a hydrogen stream and charged to the second stage.

Ⅴ. Put the following paragraphs into Chinese.

1. After the hydrocarbon leaves the first stage, it is cooled and liquefied and run through a hydrogen separator. In the hydrogen separator the hydrogen is split out for recycle to the feed. The liquid is charged to a fractionator. Whatever products are desired, the fractionator cuts them out of the first stage reactor effluent, leaving a bottom stream ready for the next, second stage, reactor.

2. In the second stage, run more severely, the catalyst opens more rings of

the heavy, complicated molecules and cracks others, forming light products. In the third stage, a kind of polishing job takes place where the olefins and aromatic compounds are saturated, forming naphthenes, paraffins, and especially isoparaffins.

4.5　Alkylation

🔲 Guidance to Reading

Alkylation is the reaction of propylene or butylene with isobutane to form an isoparaffin called alkylate. Contrary to cracking, alkylation starts with small molecules and ends up with larger ones. The alky plant consists of seven main parts. The alky plant manager has to watch a number of key variables to keep too many side reactions from occurring that could cause the quality of the alky-date to deteriorate. Alkylate has emerged as a hero to improve the environmental qualities of gasoline, since it has a low vapor pressure, zero sulfur, zero olefin content, zero benzene, and a high octane number to be a blender's dream.

🔲 Text

After the engineers were so clever about the invention of cat cracking, they attacked the problem of all the light ends the process created. The objective was to maximize the volume of gasoline being produced, but butylenes and propylene were too **volatile** and plentiful to stay **dissolved** in the gasoline blends. So they **devised** a process that was the **inverse** of cracking, alkylation, which starts with small molecules and ends up with larger ones.

The Chemical Reaction

To a chemist, alkylation can cover a broad range of reactions that stick molecules together. To a refiner, alkylation is the reaction of propylene or butylene with isobutane to form an isoparaffin called **alkylate** (Fig. 4.7).

Alkylation has a volumetric effect on refining operations the inverse of cracking because there is a significant amount of **shrinkage**. With propylene as the feed, 1 bbl of propylene and 1.6 bbl of isobutane go in and 2.1 bbl of product come out; 1 bbl of butylene and 1.21 bbl of isobutane yield 1.8 bbl of product. As in cracking, the weight in equals the weight out. Nothing gets

Isobutane
(C$_4$H$_{10}$)
Propylene
(C$_3$H$_6$)
Isoheptane
(C$_7$H$_{16}$)

Isobutane
(C$_4$H$_{10}$)
Butylene
(C$_4$H$_8$)
Isooctane
(C$_8$H$_{18}$)

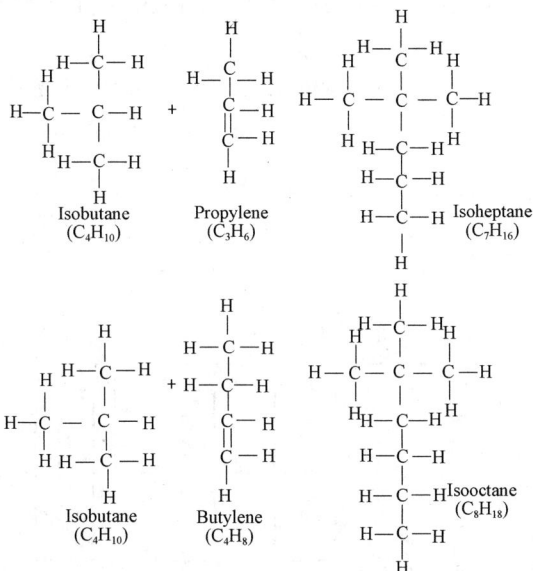

Fig. 4. 7　Alkyation of Propylene and Butylene

lost. Only the densities and volumes change.

The Process

Propylene and butylenes are **hyper** enough that the chemical reaction could be made to take place by just subjecting the isobutane and olefins, to high pressures. However the equipment would be very expensive to handle this route to alkylation. Like a lot of other processes, catalysts have been developed to facilitate the process and simplify the hardware. Alkylation plants use either sulfuric acid or hydrofluoric acid as the catalyst. In this case the catalyst is liquid, in contrast to the solids in cat cracking. The processes for both are basically the same. Those plants that use hydrofluoric acid are called HF plants; the others are called sulfuric plants. Both have safety concerns because **hydrofluoric and sulfuric acid**, are seriously **nasty** items, corroding all but the specially lined vessels and piping around them. Hydrofluoric acid has additional concern. If it escapes, it floats in a cloud and can travel great distance, to the **annoyance** of refinery neighbors. Sulfuric acid will form droplets as it escapes and quickly settles down to the ground, though that's not much **consolation** to anyone working in the immediate **vicinity**.

Sulfuric plants work better for butylene alkylation, HF plants are better for propylene. Sulfuric acid seems to be slightly more popular, so this article describes only the sulfuric route. However, the HF route to alkylation is not much different.

The alky plant consists of seven main parts: the **chiller**, the reactors, the acid separator, the caustic wash, and three distilling columns, all in, Fig. 4. 8.

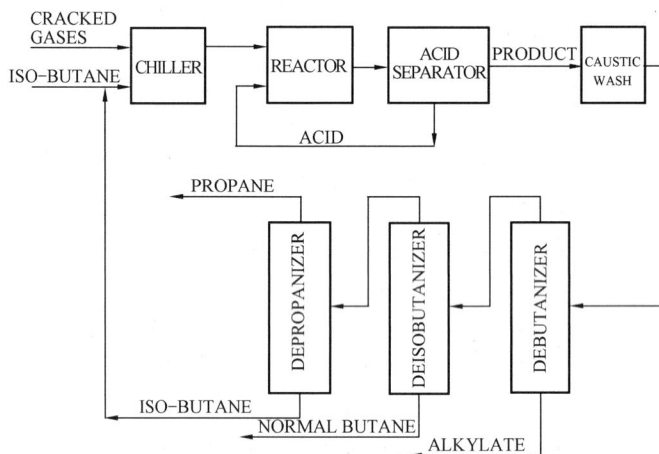

Fig. 4. 8　Alkylation Plant

The chiller. Alkylation with sulfuric acid catalysts, works best at temperatures in the neighborhood of about 40 ℉. The olefin feed (a propane/propylene or butane/butylene stream from the cracked gas plant) and a stream of isobutane is pumped through a chiller and mixed with a stream of sulfuric acid. The pressure is high enough to keep the mixture in liquid form. Sometimes the chilling is done right in the reactor.

The reactors. The reaction time for the alkylation process is relatively long, so the mixture is pumped into a battery of large reactors. The reactors hold so much total volume that by the time they all turn over once, the residence time of any one molecule is quite long, about 15 to 20 minutes. As the liquid passes through the reactors, it **encounters** mixers to assure that the olefins come in good contact with the isobutane and the acid catalyst, permitting the reaction to occur.

Acid separator. The mixture then moves to a vessel where no mixing takes place, and the acid and hydrocarbons separate like oil and water. The hydrocarbon is drawn off the top; the acid is drawn off the bottom. The acid is then recycled back to the feed side. The **acid separator** is also referred to as the acid settler.

Caustic wash. The hydrocarbon from the acid separator will have some traces of acid in it, so it is treated with caustic soda in a vessel. Caustic soda does to the hydrocarbon what Alka – Seltzer does to your stomach when you have **indigestion**—it **neutralizes** the acid. What's left (in the alky plant, not your stomach) is a mixture of hydrocarbons ready to be separated.

Fractionators. Three standard fractionators separate the alkylate and the **saturated** gas. Any unreacted isobutane is recycled to the feed.

Process Variables

The alky plant manager has to watch a number of key variables to keep too many side reactions from occurring that could cause the quality of the alkydate to **deteriorate**, as evidenced by such things as lower octane umber, poor color, and high vapor pressure.

Reaction temperature. Temperatures too low cause the sulfuric acid to get **syrupy**. That **inhibits** complete mixing and the olefins do not completely react. High temperatures cause compounds other than isoheptane and isooctane to occur, lowering the alkylate quality.

Acid strength. As the acid circulates through the process, it gets **diluted** with water that inevitably comes in with the olefins and also picks up **tar**. As the acid concentration goes from 99% down to about 89%, it is drawn off and sent back to the acid supplier for **refortify**ing.

Isobutane concentration. By having an excess amount of isobutane, the process works better. Isobutane recirculation systems are generally built in. The ratio of isobutane to olefin varies from 5:1 to 15:1. The room in the reactors usuallylimits the concentration.

Olefin space velocity. The length of time the fresh olefin feed resides in the reactor causes alkylate quality to vary.

Review

Alkylate has emerged as a hero in the past few years as refiners struggle to improve the environmental qualities of gasoline. In gasoline blending, alkylate has a low vapor pressure, zero sulfur, zero olefin content, zero benzene, and a high octane number—a blender's dream. All this comes from working with some otherwise low – valued cats and dogs around the refinery—propylene, butylene, and isobutane.

In the big refining picture the alky plant can be represented by a box with propylene, butylene, isobutane, propane, and normal butane inside. On the outside are the alkylate, propane, and normal butane. To put alkylation in perspective, Fig. 4. 9 shows the refinery processing units covered so far, plus alkylation. How complicated it quickly gets.

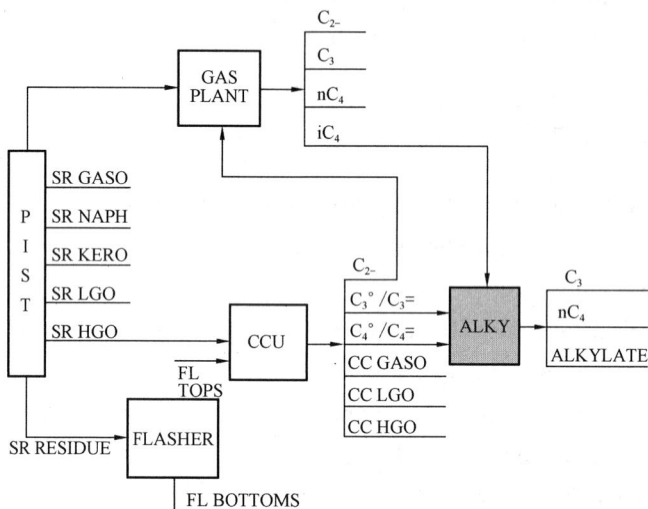

Fig. 4. 9   Refinery Proessing Operations

🔲 Words and Expressions

volatile	adj.	易挥发的
dissolve	v.	溶解
inverse	adj.	相反的
devise	v.	设计
alkylate	n.	烷化物
shrinkage	n.	收缩
hyper	adj.	超级的
nasty	adj.	有害的
annoyance	n.	讨厌的事
consolation	n.	安慰
vicinity	n.	邻居
chiller	n.	冷却器
saturate	v.	使饱和
indigestion	n.	消化不良

neutralize	v. 中和
deteriorate	v. 恶化;衰退
syrupy	adj. 糖的
dilute	v. 稀释
tar	n. 焦油
ratio	n. 比率
refortify	v. 使加固;使强化
encounter	v. 遇到
inhibit	v. 抑制

Technical Terms

hydrofluoric acid	氢氟酸
sulfuric acid	硫酸
acid separator	除酸器
caustic wash	碱洗

Language Focus

1. With propylene as the feed, 1 bbl of propylene and 1.6 bbl of isobutane go in and 2.1 bbl of product come out.

（参考译文为:当使用丙烯为原料时,1bbl 丙烯和 1.6bbl 异丁烷反应可生成 2.1 bbl 产品。）

本句中"with propylene as the feed"是独立主格结构。

2. Propylene and butylenes are hyper enough that the chemical reaction could be made to take place by just subjecting the isobutane and olefins, to high pressures.

句中"hyper"一词,意为"活跃",表示丙烯和丁烯的反应活性很高,只需将异丁烷与烯烃的混合物置于高压环境下即可发生反应。另外,句中"subject to"在科技英语中常用,在本句中表示"将……置于……条件中",在本句中表示"在高压的条件下反应可以发生"。

3. On the outside are the alkylate, propane, and normal butane.

句中"on the outside"介宾短语作状语,位于句子开头句子多用倒装形式。类似的用法有:Under the tree stood a little girl.

4. Both have safety concerns because hydrofluoric and sulfuric acid, are seriously nasty items, corroding all but the specially lined vessels and piping around them.

（参考译文为:因为氢氟酸和硫酸都具有很强的腐蚀性,能腐蚀所有的容器和管线,除非有防腐内衬。所以,使用氢氟酸和硫酸作为催化剂必须关注安全性。)

句中"concern"表示"担忧",两者都有安全方面的担忧。另外,"corro-ding all but the specially lined vessels and piping around them"在句中为现在分词短语作状语,表示修饰词的特征。

5. The alky plant consists of seven main parts: the chiller, the reactors, the acid separator, the caustic wash, and three distilling columns, all in, Fig. 4. 9.

句中"consist of"意为"由……组成",在科技英语中常用。例如:The United Kingdom consists of Great Britain and Northern Ireland. 联合王国由大不列颠和北爱尔兰组成。

6. As the liquid passes through the reactors, it encounters mixers to assure that the olefins come in good contact with the isobutane and the acid catalyst, permitting the reaction to occur.

句中"as"表示时间,"当……的时候",本句意为"当反应物流经反应器,搅拌器对其进行搅拌,使得烯烃和异丁烷与硫酸催化剂充分混合,确保反应进行"。"assure"一词表示确保的意思,后常接"that"从句,例如:I am assured that he is honest.

7. The alky plant manager has to watch a number of key variables to keep too many side reactions from occurring that could cause the quality of the alky-date to deteriorate, as evidenced by such things as lower octane umber, poor color, and high vapor pressure.

(参考译文为:烷基化副反应在很大程度上会降低烷基化油的质量,如产品辛烷值低、色泽度差和蒸汽压过大。因此,烷基化装置必须对一系列重要的参数进行控制,以抑制过多副反应的发生。)

句中"watch"意为"监控,控制",以抑制过多副反应的发生。另外,句中"evidence"作动词,表示副反应的情况确实存在。

8. Alkylate has emerged as a hero in the past few years as refiners struggle to improve the environmental qualities of gasoline.

(参考译文为:近年来,在炼油厂商为提高汽油的环境质量努力的过程中,烷基化起着极其重要的作用。)

句中"hero"原意为"英雄",本句中意为"发挥巨大作用的物",隐喻用法使语言表达更加生动。

🔲 Reinforced Learning

I. Answer the following questions for a comprehension of the text.

1. What problems did engineers try to attack after the invention of cat

cracking?

2. What is the problem with equipment engaging in subjecting the isobutene and olefins to high pressure?

3. How many parts does the alky plant consist of?

4. Why does manager have to watch variables to keep too many side reactions from occurring?

5. What plays an important role in the past few years to improve the environmental qualities of gasoline?

Ⅱ. Multiple choice：choose the correct one from the alternative answers to give the exact meaning of the word underlined.

1. So they devised a process that was the <u>inverse</u> of cracking, alkylation, which starts with small molecules and ends up with larger ones.

　A. invasion　　B. converse　　C. invention　　D. intention

2. To a chemist, alkylation can cover a <u>broad</u> range of reactions that stick molecules together.

As in cracking, the weight in equals the weight out. Nothing gets lost.

　A. plenty　　B. large　　C. wide　　D. huge

3. Like a lot of other processes, catalysts have been developed to <u>facilitate</u> the process and simplify the hardware.

　A. accelerate　　B. motivate　　C. decelerate　　D. assist

4. Sulfuric acid will form droplets it escapes and quickly settles down to the ground, though that's not much consolation to anyone working in the immediate <u>vicinity</u>.

　A. neighborhood　B. victim　　C. factory　　D. plant

5. The reaction time for the alkylation process is <u>relatively</u> long, so the mixture is pumped into a battery of large reactors.

　A. extremely　　　　　B. comparatively

　C. pretty　　　　　D. unexpectedly

6. The acid is then <u>recycled</u> back to the feed side. The acid separator is also referred to as the acid settler.

　A. reprocessed　B. discarded　　C. treasured　　D. dissolved

7. The length of time the fresh olefin feed resides in the reactor causes alkylate quality to <u>vary</u>.

　A. assess　　B. change　　C. criticize　　D. resolve

8. On the outside are the alkylate, propane, and <u>normal</u> butane.

    A. raw   B. mature   C. unusual   D. common

9. How <u>complicated</u> it quickly gets.

    A. complex   B. compound   C. comprise   D. complete

10. The length of time the fresh olefin feed <u>resides</u> in the reactor causes alkylate quality to vary.

    A. lives   B. reacts   C. stays   D. dissolves

III. Multiple choice: read the four suggested translations and choose the best answer.

1. Like a lot of other processes, catalysts have been developed to facilitate the process and <u>simplify</u> the hardware.

    A. 复杂化   B. 简化   C. 生产   D. 优化

2. So they <u>devised</u> a process that was the inverse of cracking, alkylation, which starts with small molecules and ends up with larger ones.

    A. 采用   B. 发明   C. 设计   D. 应用

3. If it escapes, it <u>floats</u> in a cloud and can travel great distance, to the annoyance of refinery neighbors.

    A. 航行   B. 飘走   C. 飞行   D. 浮动

4. The reaction time for the alkylation process is relatively long, so the <u>mixture</u> is pumped into a battery of large reactors.

    A. 合成物   B. 混合物   C. 化合物   D. 有机物

5. The hydrocarbon from the acid separator will have some <u>traces</u> of acid in it, so it is treated with caustic soda in a vessel.

    A. 脚印   B. 印记   C. 追踪   D. 寻找

IV. Put the following sentences into Chinese.

1. To a chemist, alkylation can cover a broad range of reactions that stick molecules together.

2. Alkylation has a volumetric effect on refining operations the inverse of cracking because there is a significant amount of shrinkage.

3. However the equipment would be very expensive to handle this route to alkylation.

4. Sulfuric acid will form droplets as it escapes and quickly settles down to the ground, though that's not much consolation to anyone working in the

immediate vicinity.

5. The acid is then recycled back to the feed side. The acid separator is also referred to as the acid settler.

V. Put the following paragraphs into Chinese.

1. Sulfuric plants work better for butylene alkylation, HF plants are better for propylene. Sulfuric acid seems to be slightly more popular, so this article describes only the sulfuric route. However, the HF route to alkylation is not much different.

2. Alkylate has emerged as a hero in the past few years as refiners struggle to improve the environmental qualities of gasoline. In gasoline blending, alkylate has a low vapor pressure, zero sulfur, zero olefin content, zero benzene, and a high octane number—a blender's dream. All this comes from working with some otherwise low – valued cats and dogs around the refinery—propylene, butylene, and iso – butane.

Chapter 5　Oil Refinery Products and Petrochemicals

5.1　Oil Refinery Products (1)

🔲 Guidance to Reading

In the 1860s, when modern refinery practices began, the main products from raw petroleum were lamp kerosene and residue for use as a lubricant. Nowadays however, most of all crude oil ends up as fuel of one kind or another. The fuels refined from crude oil can be divided into two general types: (1) fuels that are exploded, when vaporized with air, to provide primary moving power, and (2) those fuels that are either burned directly for heat and light or are converted into secondary energy sources such as electricity. The former type includes the aviation fuels (Avgas, Avtag, Avtur), industrial and domestic gases (LPG), motor spirits, diesel oil, and refinery gases. Fuels of the latter type include the ordinary kerosene and the various grades of fuel oils.

🔲 Text

Gasoline

Gasoline is a complex mixture of hydrocarbons that boils < 200℃ (392℉). Gasolines can vary widely in composition, even those having the same octane number can be quite different. Because of the differences in composition of the various gasolines, gasoline blending is necessary. The physical process of blending the components is simple, but determination of how much of each component to include in a blend is much more difficult. The operation is carried out by simultaneously pumping all the components of a gasoline blend into a pipeline that leads to the gasoline storage, and the pumps must be set to deliver automatically the proper proportion of each component. Sophisticated instrumentation is employed to achieve the desired blends.

The objective of gasoline (in fact, product) blending is to allocate the available blending components in such a way as to meet product demands and specifications. In the blending process, product streams from other units are collected and blended to produce the desired product. For example, in gasoline blending, product streams from **hydrotreating units**, **reforming units**, **polymerization units** and **alkylation units** are blended to produce specification gasoline.

In winter, the extreme temperatures in the northern climates allow a certain amount of butane to be **dissolved** in the gasoline to facilitate **vaporization**. The narrower boiling range ensures better distribution of the vaporized fuel through the more complicated induction systems of aircraft engines. However, **reformulation** may increase gasoline consumption, when in fact the converse is preferable. The production of low sulfur gasoline (and diesel fuel) is a necessary step to assure clean air and to reduce air pollution from the transportation sector. Refiners are using **hydrotreating technologies** in which the feedstock (be it a distillate fraction or even the residuum) is treated with hydrogen to reduce sulfur levels, and hence to make processing easier.

Kerosene

Kerosene, also called kerosine, originated as a straight run (distilled) petroleum fraction that boiled over the temperature range of 205 ~ 260℃. Kerosene was the major refinery product before the onset of the automobile age, but now kerosene might be termed as one of several other petroleum products after gasoline. In the early days of petroleum refining, some crude oils contained kerosene fractions of high quality, but other crude oils, such as those having a high proportion of asphaltic materials, had to be thoroughly refined to remove **aromatics** and sulfur compounds before a satisfactory kerosene fraction could be obtained. Kerosene is believed to be composed chiefly of hydrocarbons containing 12 to 15 carbon atoms per molecule. Low proportions of aromatic and **unsaturated hydrocarbons** are desirable to maintain the lowest possible level of smoke during burning. Although some aromatics may occur within the boiling range assigned to kerosene, excessive amounts can be removed by extraction. The significance of the total sulfur content of kerosene varies greatly with the type of oil and the use to which it is put. Sulfur content is of great importance when the kerosene to be burned produces sulfur oxides, which are of environmental concern. The color of kerosene is of little significance, but a product darker than usual may have resulted from contamination or **aging**; in fact, a color darker than specified may be considered by some users as unsatisfactory. Kerosene, because of its use as a burning oil, must be free of aromatic and unsaturated hydrocarbons; the desirable constituents of kerosene are saturated hydrocarbons. Diesel fuel, jet fuel, kerosene (range oil), No. 1 fuel oil, No. 2 fuel oil, and diesel fuel are all popular distillate products coming from the kerosene fraction of petroleum. One grade of jet fuel uses the heavy naphtha fraction, but the kerosene fraction supplies the more popular and heavier grade of jet fuel, as well as smaller amounts that are sold as burner fuel (range oil) or No. 1 heating oil. Some heating oils (generally No. 2 heating oil) and diesel fuel are similar and can sometimes substitute for each other.

Fuel Oil

Fuel oil is classified in several ways, but generally into two main types: distillate fuel oil and residual fuel oil. Distillate fuel oil is vaporized and **condensed** during a distillation process; it has a definite boiling range and does not contain high boiling oils or asphaltic components.

A fuel oil that contains any amount of the residue from crude distillation hydrocracking is a residual fuel oil. However, the terms distillate fuel oil and residual fuel oil are losing their significance because fuel oils are made for specific uses and can be either distillates, residuals, or mixtures of the two. The terms domestic fuel oil, diesel fuel oil, and heavy fuel oil are more indicative of the uses of fuel oil. Domestic fuel oils are those fuel oils used primarily in the home and include kerosene, stove oil, and furnace fuel oil. Diesel fuel oils are also distillate fuel oils, but residual oils have been successfully used to power marine diesel engines, and mixtures of distillates and residuals have been used on locomotive diesels.

Heavy fuel oils include a variety of oils, ranging from distillates to residual oils, that must be heated to 260°C or higher before they can be used. In general, **heavy fuel oil** consists of residual oil blended with distillate to suit specific needs. Heavy fuel oil includes various industrial oils and, when used to fuel ships, is called bunker oil. Stove oil is a straight-run (distilled) fraction from crude oil, whereas other fuel oils are usually blends of two or more fractions. The straight-run fractions available for blending into fuel oils are heavy naphtha, light and heavy gas oils, and residua. Cracked fractions such as light and heavy gas oils from catalytic cracking, cracking coal tar, and fractionator bottoms from catalytic cracking may also be used as blends to meet the specifications of different fuel oils. Heavy fuel oil usually contains **residuum** that is mixed to a specified viscosity with gas oils and fractionator bottoms. For some industrial purposes in which flames or **flue gases** contact the product (eg, ceramics, glass, heat treating, and **open hearth furnaces**), fuel oils must be blended to low sulfur specifications; low sulfur residues are preferable for these fuels. The manufacture of fuel oils at one time largely involved using what was left after removing desired products from crude petroleum. Now fuel oil manufacture is a complex matter of selecting and blending various petroleum fractions to meet definite specifications. Fuel oil that is used for heating is graded from No. 1 fuel oil, to No. 6 fuel oil, and cover **light distillate oils**, **medium distillate**, **heavy distillate**, a blend of distillate and residue, and residue oil.

Lubricating Oil

Lubricating oil is distinguished from other petroleum fractions by the high (>400℃) boiling point as well as their high viscosity. Lubricating oil may be divided into many categories according to the types of service; however, there are two main groups: oils used in intermittent service, such as motor and aviation oils, and oils designed for continuous service, such as turbine oils. Lubricating oil used in intermittent service must show the least possible variation in viscosity with respect to temperature and must be changed at frequent intervals to remove the foreign matter collected during service. The stability of such oil is therefore of less importance than the stability of oil used in continuous service for prolonged periods without renewal. Lubricating oil for continuous service must be extremely stable because the engines in which it is used operate at fairly constant temperature without frequent shutdown.

Words and Expressions

dissolved	adj. 溶解的
vaporization	n. 蒸发
reformulation	n. 新配方汽油
aromatics	n. 芳烃
aging	adj. 老化的
condensed	adj. 浓缩的

residuum	n. 残渣,渣油

Technical Terms

hydrotreating unit	加氢精制装置
reforming unit	重整装置
polymerization unit	叠合装置
alkylation unit	烷基化装置
hydrotreating technology	加氢技术
unsaturated hydrocarbon	不饱和烃
heavy fuel oil	重质燃料油
flue gases	烟道气
open hearth furnace	平炉
light distillate oil	轻质馏分油
medium distillate	中质馏分
heavy distillate	重质馏分

Language Focus

1. The objective of gasoline (in fact, product) blending is to allocate the available blending components in such a way as to meet product demands and specifications.

[参考译文为:汽油(实际是产品)混兑的目的是调配现有组分以满足产品的需要和规格。]

本句为科技类文章中常用的句型:the objective of sth. is to do sth.,句中有两个目的,其中前一个目的又是后一个目的的条件。"so/such...as to"与"so/such...that"同义,引导结果或目的状语从句,词组"so as to do sth."是其变形,意为"以便……,为的是……",例如:He shouted and waved so as to be noticed. 他又喊又招手以引起别人的注意。

2. In winter, the extreme temperatures in the northern climates allow a certain amount of butane to be dissolved in the gasoline to facilitate vaporization.

(参考译文为:北方冬季气温极低,因而汽油里可以有一定量的丁烷,以利于汽油蒸发。)

句中主语为"temperatures",谓语动词是"allow",为拟人化手法,表达生动形象。"the extreme temperatures in winter"实际指的就是北方冬季的严寒天气,翻译时可直接译为"冬天北方的严寒天气……"。

3. Kerosene, also called kerosine, originated as a straight run (distilled) petroleum fraction that boiled over the temperature range of 205 ~ 260℃.

[参考译文为:煤油,也称为"kerosine",是石油在 205 ~ 260℃ 的温度范围内产生的直馏(蒸馏)馏分。]

句中"that"引导定语从句,修饰先行词"fraction"。

4. Kerosene, because of its use as a burning oil, must be free of aromatic and unsaturated hydrocarbons; the desirable constituents of kerosene are saturated hydrocarbons.

(参考译文为:煤油作为燃料使用,所以不能含有芳烃与不饱和烃。煤油的理想成分是饱和烃。)

句中"be free of"表示"不含有",在科技英语中常用。

5. Fuel oil that is used for heating is graded from No. 1 fuel oil, to No. 6 fuel oil, and cover light distillate oils, medium distillate, heavy distillate, a blend of distillate and residue, and residue oil.

(参考译文为:用于加热的燃料油分 1 号 ~ 6 号 6 个等级,包括轻馏分油、中度馏分油、重馏分油、馏分油与残渣的混合组分以及渣油。)

句中,"is graded from No. 1 fuel oil, to …"表示"等级范围从……到……"在科技英语中常用。

Reinforced Learning

Ⅰ. **Answer the following questions for a comprehension of the text.**

1. What's the objective of gasoline blending?

2. What's the significance for low sulfur gasoline?

3. How can we get desirable kerosene from crude oils with a high proportion of asphaltic materials?

4. What are the two main types in fuel oil?

5. How can we distinguish lubricating oil from others?

Ⅱ. **Multiple choice: choose the correct one from the alternative answers to give the exact meaning of the word underlined.**

1. The objective of gasoline (in fact, product) blending is to allocate the available blending components in such a way as to meet product demands and specifications.

A. assign   B. distribute

C. set   D. allot

2. The narrower boiling range ensures better distribution of the vaporized fuel through the more complicated <u>induction</u> systems of aircraft engines.

A. introduction

B. enlistment

C. the process by which electrical or magnetic properties are transferred

D. reasoning by which known facts

3. Refiners are using hydrotreating technologies in which the feedstock (be it a distillate fraction or even the residuum) is treated with hydrogen to reduce sulfur <u>levels</u>, and hence to make processing easier.

A. height   B. quantity   C. quality   D. position

4. Kerosene was the major refinery product before the <u>onset</u> of the automobile age, but now kerosene might be termed as one of several other petroleum products after gasoline.

A. beginning   B. setting

C. playing   D. processing

5. Low proportions of aromatic and unsaturated hydrocarbons are <u>desirable</u> to maintain the lowest possible level of smoke during burning.

A. attractive   B. reasonable

C. satisfactory   D. expected

6. Diesel fuel, jet fuel, kerosene (range oil), No. 1 fuel oil, No. 2 fuel oil, and diesel fuel are all <u>popular</u> distillate products coming from the kerosene fraction of petroleum.

A. famous   B. main   C. common   D. public

7. Diesel fuel oils are also distillate fuel oils, but residual oils have been successfully used to <u>power</u> marine diesel engines, and mixtures of distillates and residuals have been used on locomotive diesels.

A. force   B. push   C. start   D. motivate

8. The manufacture of fuel oils at one time <u>largely</u> involved using what was left after removing desired products from crude petroleum.

A. chiefly   B. deeply

C. impressively   D. partly

9. Fuel oil that is used for heating is graded from No. 1 fuel oil, to No. 6

fuel oil, and <u>cover</u> light distillate oils, medium distillate, heavy distillate, a blend of distillate and residue, and residue oil.

    A. include   B. spread   C. take up   D. protect

10. Lubricating oil used in intermittent service must show the least possible variation in viscosity with respect to temperature and must be changed at frequent intervals to remove the <u>foreign</u> matter collected during service.

    A. aboard   B. unfamiliar

    C. outside   D. outward

Ⅲ. Multiple choice：read the four suggested translations and choose the best answer.

1. The operation is carried out by simultaneously pumping all the components of a gasoline blend into a pipeline that <u>leads to</u> the gasoline storage, and the pumps must be set to deliver automatically the proper proportion of each component.

    A. 导致   B. 输送   C. 引导   D. 领先

2. Although some aromatics may occur within the boiling range <u>assigned to</u> kerosene, excessive amounts can be removed by extraction.

    A. 指派   B. 指定   C. 涉及   D. 属于

3. Kerosene, because of its use as a burning oil, must <u>be free of</u> aromatic and unsaturated hydrocarbons; the desirable constituents of kerosene are saturated hydrocarbons.

    A. 自由做……   B. 免费做……

    C. 随便做……   D. 免于,免除

4. The terms domestic fuel oil, diesel fuel oil, and heavy fuel oil are more <u>indicative of</u> the uses of fuel oil.

    A. 表明   B. 显示   C. 介绍   D. 描述

5. Lubricating oil used in intermittent service must show the least possible variation in viscosity <u>with respect</u> to temperature and must be changed at frequent intervals to remove the foreign matter collected during service.

    A. 考虑到   B. 关于

    C. 在……方面   D. 受制于

Ⅳ. Put the following sentences into Chinese.

1. Gasoline is a complex mixture of hydrocarbons that boils $< 200\,^\circ\!C$

(392 ℉).

2. The production of low sulfur gasoline (and diesel fuel) is a necessary step to assure clean air and to reduce air pollution from the transportation sector.

3. However, reformulation may increase gasoline consumption, when in fact the converse is preferable.

4. Kerosene was the major refinery product before the onset of the automobile age, but now kerosene might be termed as one of several other petroleum products after gasoline.

5. Sulfur content is of great importance when the kerosene to be burned produces sulfur oxides, which are of environmental concern.

V. Put the following paragraphs into Chinese.

1. The objective of gasoline (in fact, product) blending is to allocate the available blending components in such a way as to meet product demands and specifications. In the blending process, product streams from other units are collected and blended to produce the desired product. For example, in gasoline blending, product streams from hydrotreating units, reforming units, polymerization units and alkylation units are blended to produce specification gasoline.

2. Fuel oil is classified in several ways, but generally into two main types: distillate fuel oil and residual fuel oil. Distillate fuel oil is vaporized and condensed during a distillation process; it has a definite boiling range and does not contain high boiling oils or asphaltic components.

5.2   Oil Refinery Products (2)

⌐⊓ Guidance to Reading

Nowadays, about 88% of all crude oil ends up as fuel of one kind or another. Of the remaining 12% , just over half is refined into petroleum – chemical intermediates. These are used as feedstocks in the manufacture of synthetic materials (fibres, rubbers, plastics, etc.), fertilizers, insecticides, and even protein for animal feeds. About 5% of the average barrel of crude is used in the production of a wide range of lubricating oils and greases, waxes, solvents, and asphalt for roads and weatherproofing.

⬛ **Text**

Wax

Petroleum waxes are of two general types: paraffin wax in distillates and **microcrystalline wax** in residua. The melting point of wax is not directly related to its boiling point because waxes contain hydrocarbons of different chemical structure. Nevertheless, waxes are graded according to their melting point and oil content. **Paraffin wax** is a solid **crystalline** mixture of **straight – chain** (normal) hydrocarbons ranging from mostly 20 ~ 30 and higher. Wax constituents are solid at ordinary temperatures (25℃; 77℉) whereas **petrolatum** (**petroleum jelly**) contains both solid and liquid hydrocarbons. Wax production by **wax sweating** was originally used in Scotland to separate wax fractions by employing various melting points from the wax obtained from shale oils. Wax sweating is still used to some extent, but is being replaced by the more convenient wax **recrystallization** process. In wax sweating, a cake of **slack wax**, also known as **crude** or **raw wax**, is slowly warmed to a temperature at which the oil in the wax and the lower melting waxes become fluid and drip (or sweat) from the bottom of the cake, leaving a residue of higher melting wax.

Insofar as they are used to purify other products, several processes used in the refinery fall under the classification of **dewaxing** processes; however, such processes must also be classified as wax production processes. Most commercial dewaxing processes utilize **solvent dilution**, chilling to **crystallize** the wax, and **filtration**. The MEK process (MEKoluene solvent) is widely used. **Wax crystals** are formed by chilling through the walls of **scraped surface chillers**, and wax is separated from the resultant **wax solvent slurry** by using fully enclosed rotary vacuum filters.

On the other hand, intermediate paraffin distillates contain paraffin waxes and waxes intermediate in properties between paraffin and microwaxes. Thus, the solvent dewaxing process produces three different slack waxes depending on whether light, intermediate, or heavy paraffin distillate is processed. The slack wax from heavy paraffin distillate may be sold as dark raw wax, the wax from intermediate paraffin distillate as pale raw wax. The latter is treated with **lye** and clay to remove odor and improve color.

In the propane process, part of the propane diluent is allowed to evaporate

by reducing pressure so as to chill the **slurry** to the desired filtration temperature, and rotary pressure filters are employed. Complex dewaxing requires no
refrigeration, but depends on the formation of a solid urea – paraffin complex
that is separated by filtration and then decomposed. This process is used to
make low viscosity lubricants that must remain fluid at low temperatures (refrigeration, transformer, and **hydraulic oils**).

Catalytic dewaxing processes based on selective hydrocracking of the **normal paraffins** use a molecular sieve – based catalyst in which the active hydrocracking sites are accessible only to the paraffin molecules. **Catalytic dewaxing**
is a hydrocracking process operated at elevated temperatures ($280 \sim 400 °C$; 536
$\sim 752 °F$) and pressures, $2070 \sim 3450 kPa(300 \sim 500 psi)$. However, the conditions for a specific dewaxing operation depend on the nature of the feedstock
and the product **pour point** required. The catalyst employed for the process is a
mordenite – type catalyst that has the correct pore structure to be selective for
normal paraffin cracking. Platinum on the catalyst serves to hydrogenate the **reactive intermediates** so that further paraffin degradation is limited to the initial
thermal reactions.

Another catalytic dewaxing process also involves selective cracking of normal paraffins and those paraffins that may have minor branching in the chain.
In the process, the catalyst can be reactivated to fresh activity by relatively mild
nonoxidative treatment. The time allowed between reactivations is a function of
the feedstock; after numerous reactivations it is possible that there will be coke
buildup on the catalyst. A catalytic dewaxing process can be used to dewax a
variety of lubricating base stocks; as such, it has the potential to replace solvent
dewaxing, or even be used in combination with solvent dewaxing, as a means
of relieving the bottlenecks which can, and often do, occur in solvent dewaxing
facilities.

Asphalt

Asphalt manufacture is, in essence, a matter of distilling everything possible from crude petroleum until a residue with the desired properties is obtained.
This is usually done by stages; crude distillation at atmospheric pressure removes the lower boiling fractions and yields a **reduced crude** that may contain
higher boiling (lubricating) oils, asphalt, and even wax. Distillation of the reduced crude under vacuum removes the oils (and wax) as volatile overhead

products and the asphalt remains as a bottom (or residual) product. At this stage the asphalt is frequently (and incorrectly) referred to as **pitch**. In terms of meeting specifications, asphalt can be made softer by blending hard asphalt with the extract obtained in the solvent treatment of lubricating oils. On the other hand, soft asphalts can be converted into harder asphalts by oxidation (air blowing).

Cutback asphalts are mixtures in which hard asphalt has been diluted with a lighter oil to permit application as a liquid without drastic heating. They are classified as rapid, medium, and slow curing, depending on the volatility of the diluent, which governs the rates of evaporation and consequent hardening. Asphalt can be **emulsified** with water to permit application without heating. Such emulsions are normally of the oil – in – water type. They reverse or break on application to a stone or earth surface, so that the oil clings to the stone and the water disappears. In addition to their usefulness in road and soil stabilizations, they are useful for **paper impregnation** and waterproofing. The emulsions are chiefly either the soap or alkaline type, or the neutral or clay type. The former breaks readily on contact, but the latter is more stable and probably loses water mainly by evaporation. Good emulsions must be stable during storage or freezing, suitably fluid, and amenable to control for the speed of breaking.

Recently, asphalt has grown to be a valuable refinery product. In the post – 1980 period, a shortage of good quality asphalt has developed. This is due in no short measure to the tendency of refineries to produce as much liquid fuels (eg, gasoline) as possible. Thus, residua that would have once been used for asphalt manufacture are now being used to produce liquid fuels (and coke).

Coke

Coke is the residue left by the **destructive distillation** (coking) of residua. The composition of coke varies with the source of the crude oil, but in general, large amounts of high molecular weight complex hydrocarbons (rich in carbon, but correspondingly poor in hydrogen) make up a high proportion. The **solubility** of coke in **carbon disulfide** has been reported to be as high as 50% ~ 80%, but this is, in fact, a misnomer, since the coke is an **insoluble**, **honeycomb – type** material that is the end product of thermal processes. Petroleum coke is employed for a number of purposes; its principal use is in the manufacture of **carbon electrodes** for aluminum refining, which requires a high purity carbon that is low in ash and free of sulfur. In addition, coke is employed in the manufacture of carbon brushes, **silicon carbide abrasives**, structural carbon (eg, pipes and rashing rings), as well as **calcium carbide** manufacture from which **acetylene** is produced. Coke produced from low quality crude oil is solvent mixed with coal and burned as a fuel. Flue gas scrubbing is required. Coke is used in **fluidized – bed combustors** or **gasifiers** for power generation.

🔲 Words and Expressions

crystalline	adj.	结晶的
recrystallization	n.	再结晶
dewaxing	n.	脱蜡
crystallize	v.	使……结晶
filtration	n.	过滤
lye	n.	碱水
slurry	n.	浆
mordenite	n.	丝光沸石
pitch	n.	沥青
emulsify	v.	乳化
solubility	n.	溶解度
insoluble	adj.	不可溶的
honeycomb – type	adj.	蜂窝状的
abrasives	n.	磨料
acetylene	n.	乙炔
gasifier	n.	气化炉

🔲 Technical Terms

paraffin wax	石蜡
normal paraffin	正构烷烃
microcrystalline wax	微晶蜡
straight – chain hydrocarbons	直链烃
petrolatum/petroleum jelly	石油蜡,凡士林
wax sweating	结晶蜡脱油
slack wax	含油蜡

crude／raw wax	粗蜡
solvent dilution	稀释溶剂
wax crystal	蜡结晶
scraped surface chiller	刮板式表面结晶器
solvent slurry	熔浆
hydraulic oil	液压油
catalytic dewaxing	催化脱蜡
pour point	倾点
reactive intermediates	活泼中间体
reduced crude	拔头原油
cutback asphalt	稀释沥青
paper impregnation	造纸
destructive distillation	干馏
carbon disulfide	二硫化碳
carbon electrode	碳电极
silicon carbide	碳化硅
calcium carbide	碳化钙
fluidized – bed combustor	流化床燃烧器

Language Focus

1. In wax sweating, a cake of slack wax, also known as crude or raw wax, is slowly warmed to a temperature at which the oil in the wax and the lower melting waxes become fluid and drip (or sweat) from the bottom of the cake, leaving a residue of higher melting wax.

（参考译文为：在结晶蜡脱油中，混油石蜡，也就是粗蜡或原料蜡，缓慢加热到一定温度，蜡和较低熔点的蜡中的油会变成液体，从结块底部滴下，留下高熔点的蜡残渣。）

句中"at which"引导定语从句修饰先行词"temperature"。

2. In the propane process, part of the propane diluent is allowed to evaporate by reducing pressure so as to chill the slurry to the desired filtration temperature, and rotary pressure filters are employed.

（参考译文为：在丙烷生产工艺中，使用旋转式压力过滤器，降低压力，使部分丙烷稀释剂得以蒸发，这样泥浆就可以达到合适的过滤温度。）

"so as to"引导目的状语，表示要让泥浆达到设定的温度。

3. A catalytic dewaxing process can be used to dewax a variety of lubricating base stocks; as such, it has the potential to replace solvent dewaxing, or even be used in combination with solvent dewaxing, as a means of relieving the bottlenecks which can, and often do, occur in solvent dewaxing facilities.

本句中"as such, it has the potential to replace solvent dewaxing"中主语"it"指代"a catalytic dewaxing process",形式主语句在科技英语中较为常用,起到调节句子平衡的作用。"bottleneck"一词本意为"瓶颈",此处引申为"困难",属于隐喻修辞。

4. In terms of meeting specifications, asphalt can be made softer by blending hard asphalt with the extract obtained in the solvent treatment of lubricating oils.

(参考译文为:为了满足产品规格,润滑油经溶剂处理后添加到硬沥青中,可使沥青变软。)

本句中"in terms of"意为"根据",在科技英语中常用。

5. They are classified as rapid, medium, and slow curing, depending on the volatility of the diluent, which governs the rates of evaporation and consequent hardening.

(参考译文为:所以根据稀释液的挥发性可将稀释沥青分为快凝、中凝、慢凝稀释沥青。)

词组"classify... among/as sth."意为"把……归为……类",例如:The film is classified as mere entertainment. 这部电影可归为纯娱乐片。

🔳 Reinforced Learning

Ⅰ. Answer the following questions for a comprehension of the text.

1. How does wax sweating happen?

2. What are the features of mordenite – type catalyst?

3. What's the essence of asphalt manufacture?

4. What are the stages of asphalt manufacture?

5. Is it right to say that coke can be soluble in carbon disulfide, and why?

Ⅱ. Multiple choice: choose the correct one from the alternative answers to give the exact meaning of the word underlined.

1. Nevertheless, waxes are graded according to their melting point and oil content.

A. classified   B. recorded

C. scored   D. marked

2. In wax sweating, a <u>cake</u> of slack wax, also known as crude or raw wax, is slowly warmed to a temperature at which the oil in the wax and the lower melting waxes become fluid and drip (or sweat) from the bottom of the cake, leaving a residue of higher melting wax.

A. shape   B. block

C. food   D. piece

3. Wax crystals are formed by chilling through the walls of scraped surface chillers, and wax is separated from the <u>resultant</u> wax solvent slurry by using fully enclosed rotary vacuum filters.

A. resulting   B. combined

C. existing   D. original

4. Catalytic dewaxing is a hydrocracking process operated at <u>elevated</u> temperatures ($280 \sim 400\,^{\circ}\text{C}$; $536 \sim 752\,^{\circ}\text{F}$) and pressures, $2070 \sim 3450\text{kPa}$($300 \sim 500\text{psi}$).

A. lifted   B. high

C. raised   D. low

5. The catalyst employed for the process is a mordenite – type catalyst that has the <u>correct</u> pore structure to be selective for normal paraffin cracking.

A. right   B. appropriate

C. numerous   D. expanded

6. Platinum on the catalyst serves to hydrogenate the reactive intermediates so that further paraffin degradation is limited to the <u>initial</u> thermal reactions.

A. beginning   B. primary

C. low – level   D. major

7. Another catalytic dewaxing process also <u>involves</u> selective cracking of normal paraffins and those paraffins that may have minor branching in the chain.

A. includes   B. requires

C. concerns   D. affects

8. Cutback asphalts are mixtures in which hard asphalt has been diluted with a lighter oil to <u>permit</u> application as a liquid without drastic heating.

A. allow   B. suggest

C. order   D. decide

9. Good emulsions must be stable during storage or freezing, suitably fluid, and <u>amenable</u> to control for the speed of breaking.

A. appropriate　　　　　　　B. equally

C. similarly　　　　　　　　D. differently

10. Flue gas <u>scrubbing</u> is required.

A. cleaning　　　　　　　　B. purifying

C. eliminating　　　　　　　D. detecting

Ⅲ. Multiple choice : read the four suggested translations and choose the best answer.

1. The slack wax from heavy paraffin distillate may be sold as dark raw wax, the wax from <u>intermediate</u> paraffin distillate as pale raw wax.

A. 中间的　　　　　　　　　B. 中质的

C. 中等的　　　　　　　　　D. 中级的

2. Catalytic dewaxing processes <u>based on</u> selective hydrocracking of the normal paraffins use a molecular sieve – based catalyst in which the active hydrocracking sites are accessible only to the paraffin molecules.

A. 基于　　　　　　　　　　B. 始于

C. 高于　　　　　　　　　　D. 产生于

3. The time allowed between reactivations is a function of the feedstock; after numerous reactivations it is possible that there will be coke <u>buildup on</u> the catalyst.

A. 建立于　　　　　　　　　B. 产生于

C. 附着于　　　　　　　　　D. 发生于

4. At this stage the asphalt is frequently (and incorrectly) <u>referred to as</u> pitch.

A. 称为　　　　　　　　　　B. 指示

C. 所指　　　　　　　　　　D. 涉及

5. They reverse or break on application to a stone or earth surface, so that the oil <u>clings to</u> the stone and the water disappears.

A. 黏附　　　　　　　　　　B. 靠近

C. 挨着　　　　　　　　　　D. 坚持

Ⅳ. Put the following sentences into Chinese.

1. The melting point of wax is not directly related to its boiling point be-

cause waxes contain hydrocarbons of different chemical structure.

2. On the other hand, intermediate paraffin distillates contain paraffin waxes and waxes intermediate in properties between paraffin and microwaxes.

3. Another catalytic dewaxing process also involves selective cracking of normal paraffins and those paraffins that may have minor branching in the chain.

4. Cutback asphalts are mixtures in which hard asphalt has been diluted with a lighter oil to permit application as a liquid without drastic heating.

5. Coke is the residue left by the destructive distillation (coking) of residua.

V. Put the following paragraphs into Chinese.

1. Insofar as they are used to purify other products, several processes used in the refinery fall under the classification of dewaxing processes; however, such processes must also be classified as wax production processes. Most commercial dewaxing processes utilize solvent dilution, chilling to crystallize the wax, and filtration. The MEK process (MEKoluene solvent) is widely used. Wax crystals are formed by chilling through the walls of scraped surface chillers, and wax is separated from the resultant wax solvent slurry by using fully enclosed rotary vacuum filters.

2. Recently, asphalt has grown to be a valuable refinery product. In the post–1980 period, a shortage of good quality asphalt has developed. This is due in no short measure to the tendency of refineries to produce as much liquid fuels (eg, gasoline) as possible. Thus, residua that would have once been used for asphalt manufacture are now being used to produce liquid fuels (and coke).

5.3 Petrochemicals

Guidance to Reading

Petrochemicals are generally chemical compounds derived from petroleum either by direct manufacture or by indirect manufacture as by – products from the variety of processes that are used during the refining of petroleum. The classification of petrochemicals is to indicate the source of the chemical compounds, and the terminology is therefore a matter of source identification. The manufacture of chemicals from petroleum is based on the ready response of the various compounds types to basic chemical reactions, such as oxidation, halo-

genation, nitration, dehydrogenation, polymerization, and alkylation. The low
− molecular − weight paraffins and olefins, as found in natural gas and refinery
gases, and the simple aromatic hydrocarbons have so far been of the most in-
terest because it is individual species that can readily be isolated and dealt
with.

🔲 Text

Petrochemicals

The petrochemical industry began in the 1920s as suitable by − products became available through improvements in the refining processes. It developed
parallel with the oil industry and has rapidly expanded since the 1940s, with the
oil refining industry providing plentiful cheap raw materials. A petrochemical is
any chemical (as distinct from fuels and petroleum products) manufactured
from petroleum (and natural gas) and used for a variety of commercial purposes. The definition, however, has been broadened to include the whole range
of **aliphatic**, aromatic, and **naphthenic** organic chemicals, as well as carbon
black and such inorganic materials as sulfur and ammonia. Petroleum and natural gas are made up of hydrocarbon molecules, which are comprised of one or
more carbon atoms, to which hydrogen atoms are attached. Currently, oil and
gas are the main sources of the raw materials because they are the least expensive, most readily available, and can be processed most easily into the primary
petrochemicals. Primary petrochemicals include olefins (ethylene, propylene
and butadiene), aromatics (benzene, toluene, and the isomers of xylene) and
methanol. Thus, petrochemical feedstocks can be classified into three general
groups: olefins, aromatics, and methanol; a fourth group includes inorganic
compounds and synthesis gas (mixtures of carbon monoxide and hydrogen). In
many instances, a specific chemical included among the petrochemicals may also be obtained from other sources, such as coal, coke, or vegetable products.
For example, materials such as benzene and **naphthalene** can be made from either petroleum or coal, while **ethyl alcohol** may be of petrochemical or vegetable origin.

Petrochemicals are generally divided into three groups: (1) aliphatics,
such as butane and butene; (2) cycloaliphatics, such as cyclohexane, cyclohexane **derivatives**, and aromatics (eg, benzene, toluene, xylene, and naph-

thalene); and (3) inorganics, such as sulfur, ammonia, **ammonium sulfate**, **ammonium nitrate**, and **nitric acid**.

Aliphatics

Methane, obtained from crude oil or natural gas, or as a product from various conversion (cracking) processes, is an important source of raw materials for aliphatic petrochemicals. Ethane, also available from natural gas and cracking processes, is an important source of ethylene, which, in turn, provides more valuable routes to petrochemical products.

Ethylene, an important olefin, is usually made by cracking gases such as ethane, propane, butane, or a mixture of these as might exist in a refinery's off – gases. When gas feedstock is scarce or expensive, naphthas and even whole crude oil have been used in specially designed ethylene **crackers**. The heavier feeds also give significant quantities of higher molecular weight olefins and aromatics. Ethylene is consumed in larger amounts than any other hydrocarbon for the production of aliphatic petrochemicals, but it is by no means the only source of aliphatic petrochemicals. Propane and butane are also important aliphatic hydrocarbons. Propane is usually converted to propylene by thermal cracking, although some propylene is also available from **refinery gas** streams. The various butylenes are more commonly obtained from refinery gas streams. Butane dehydrogenation to butylene is known, but is more complex than ethane or propane cracking, and its product distributions are not always favorable.

The production of gasoline and other liquid fuels consumes large amounts of butane. The gaseous constituents produced in a refinery give rise to a host of chemical intermediates that can be used for the manufacture of a wide variety of products. Synthesis gas (carbon monoxide, CO, and hydrogen, H_2) mixtures are also used to produce valuable industrial chemicals.

Cycloaliphatics and Aromatics

Cyclic compounds (cyclohexane and benzene) are also important sources of petrochemical products. Aromatics are in high concentration in the product streams from a **catalytic reformer**. When aromatics are needed for petrochemical manufacture, they are extracted from the reformer's product using solvents, such as **glycols** (eg, the **Udex process**) and sulfolane.

The mixed monocyclic aromatics are called BTX as an abbreviation for benzene, toluene, and xylene. The benzene and toluene are isolated by distillation, and the isomers of the xylene are separated by **superfractionation**, **fractional crystallization**, or **adsorption**. Benzene is the starting material for styrene, phenol, and a number of fibers and plastics. Benzene and cyclohexane are responsible for products such as nylon and **polyester fibers**, **polystyrene**, epoxy resins, **phenolic resins**, and **polyurethanes**.

Toluene is used to make a number of chemicals, but most of it is blended into gasoline. Xylene use depends on the isomer: **p – xylene** goes into polyester and **o – xylene** into **phthalic anhydride**. Both are involved in a wide variety of consumer products.

Benzene, toluene, and xylene are made mostly from catalytic reforming of naphthas. As a gross mixture, these aromatics are the backbone of gasoline blending for high octane numbers. However, there are many chemicals derived from these same aromatics; thus many aromatic petrochemicals have their beginning by selective extraction from naphtha or **gas – oil** reformate.

Inorganics

Of the inorganic chemicals, ammonia is by far the most common. Ammonia is produced by the direct reaction of hydrogen with nitrogen; air is the source of nitrogen. Refinery gases, steam reforming of natural gas (methane) and naphtha streams, and partial oxidation of hydrocarbons or higher molecular weight refinery residual materials (residua, asphalts) are the sources of hydrogen. Ammonia is used predominantly for the production of ammonium nitrate (NH_4NO_3), as well as other **ammonium salts** and **urea**(H_2NCONH_2), which are primary constituents of fertilizers.

Carbon black, also classed as an inorganic petrochemical, is made predominantly by the partial combustion of carbonaceous (organic) material in a limited supply of air. **Carbonaceous** sources vary from methane to aromatic petroleum oils to coal tar by – products. Carbon black is used primarily for the production of synthetic rubber. Sulfur, another inorganic petrochemical, is obtained by the oxidation of **hydrogen sulfide**: $2H_2S + O_2 = 2H_2O + 2S$.

Hydrogen sulfide is a constituent of natural gas and also of the majority of refinery gas streams, especially those off – gases from **hydrodesulfurization** processes. A majority of the sulfur is converted to sulfuric acid for the manufacture of fertilizers and other chemicals. Other uses for sulfur include the production of carbon disulfide, refined sulfur, and **pulp** and paper industry chemicals.

Words and Expressions

petrochemicals	n. 石油化工产品
aliphatic	n. 脂肪类
naphthenic	n. 环烷烃类
naphthalene	n. 环烷烃
derivative	n. 衍生物
cracker	n. 裂解炉
glycol	n. 乙二醇
superfractionation	n. 超精馏
adsorption	n. 吸附
polystyrene	n. 聚苯乙烯
polyurethane	n. 聚氨酯
p – xylene	n. 对二甲苯
o – xylene	n. 邻二甲苯
gas – oil	n. 瓦斯油
urea	n. 尿素
carbonaceous	adj. 含碳的
hydrodesulfurization	n. 加氢脱硫
pulp	n. 纸浆

Technical Terms

ethyl alcohol	乙醇
ammonium sulfate	硫酸铵
ammonium nitrate	硝酸铵
nitric acid	硝酸
refinery gas	炼厂气
catalytic reformer	催化重整器
Udex process	尤狄克斯法萃取工艺
fractional crystallization	分步结晶
polyester fibers	聚酯纤维
phenolic resins	酚醛树脂
phthalic anhydride	邻二甲酸酐
ammonium salt	铵盐

| hydrogen sulfide | 硫化氢 |

⊡ Language Focus

1. Ethylene, an important olefin, is usually made by cracking gases such as ethane, propane, butane, or a mixture of these as might exist in a refinery's off – gases.

[参考译文为:乙烯是一种重要的烯烃,通常由裂化气(如乙烷、丙烷、丁烷,或炼厂废气,即这些气体的混合物)制得。]

句中"an important olefin",意为"一种重要的烯烃",为插入语,起补充说明作用。

2. Ethylene is consumed in larger amounts than any other hydrocarbon for the production of aliphatic petrochemicals, but it is by no means the only source of aliphatic petrochemicals.

句中"by no means"表达强烈的语气,该词组意思为"决不,并没有",如:I will by no means be a professor. 我决不会成为一名教授的。

3. Butane dehydrogenation to butylene is known, but is more complex than ethane or propane cracking, and its product distributions are not always favorable.

句中"favorable"一词意为"有利的,良好的",本句中表示:该方法的产物分布也不是很尽如人意。

4. Ammonia is used predominantly for the production of ammonium nitrate (NH_4NO_3), as well as other ammonium salts and urea(H_2NCONH_2), which are primary constituents of fertilizers.

[参考译文为:氨气主要用于生产硝酸铵(NH_4NO_3)以及其他铵盐和尿素(H_2NCONH_2),这些都是化肥的主要成分。]

句中"which"引导非限定性定语从句,代替前面的"ammonium nitrate (NH_4NO_3), as well as other ammonium salts and urea(H_2NCONH_2)"。

⊡ Reinforced Learning

I. Answer the following questions for a comprehension of the text.

1 What's the broaden definition of petrochemicals?

2. What are the general three groups of petrochemicals?

3. What roles do methane, ethane and ethylene play in petrochemicals production?

4. What's the use of xylene?

5. How can we get the carbon black and what's its function?

Ⅱ. Multiple choice: choose the correct one from the alternative answers to give the exact meaning of the word underlined.

1. Currently, oil and gas are the main sources of the raw materials because they are the least expensive, most <u>readily</u> available, and can be processed most easily into the primary petrochemicals.

    A. willing   B. prepared   C. likely   D. skillfully

2. <u>Primary</u> petrochemicals include olefins (ethylene, propylene and butadiene) aromatics (benzene, toluene, and the isomers of xylene); and methanol.

    A. Simple   B. Original   C. Main   D. Elementary

3. In many <u>instances</u>, a specific chemical included among the petrochemicals may also be obtained from other sources, such as coal, coke, or vegetable products.

    A. cases   B. examples   C. events   D. times

4. Ethane, also available from natural gas and cracking processes, is an important source of ethylene, which, in turn, provides more valuable <u>routes</u> to petrochemical products.

    A. ways   B. possibilities   C. directions   D. lines

5. Butane dehydrogenation to butylene is known, but is more complex than ethane or propane cracking, and its product <u>distributions</u> are not always favorable.

    A. divisions   B. sales   C. spreads   D. productions

6. Butane dehydrogenation to butylene is known, but is more complex than ethane or propane cracking, and its product distributions are not always <u>favorable</u>.

    A. satisfactory   B. approval

    C. approachable   D. achievable

7. The gaseous constituents produced in a refinery give rise to a host of chemical intermediates that can be used for the manufacture of a <u>wide</u> variety of products.

    A. large   B. landing   C. broad   D. loose

8. Aromatics are in high <u>concentration</u> in the product streams from a catalytic reformer.

A. strength   B. density   C. proportion   D. attention

9. The benzene and toluene are <u>isolated</u> by distillation, and the isomers of the xylene are separated by superfractionation, fractional crystallization, or adsorption.

A. separated   B. alienate   C. set apart   D. cut off

10. As a gross mixture, these aromatics are the <u>backbone</u> of gasoline blending for high octane numbers.

A. firmness   B. strength   C. support   D. basis

III. Multiple choice : read the four suggested translations and choose the best answer.

1. The petrochemical industry began in the 1920s as <u>suitable</u> by – products became available through improvements in the refining processes.

A. 适合的   B. 适当的   C. 相配的   D. 适宜的

2. Ethane, also available from natural gas and cracking processes, is an important source of ethylene, which, <u>in turn</u>, provides more valuable routes to petrochemical products.

A. 反过来   B. 轮流   C. 按顺序   D. 转变

3. Benzene is the <u>starting material</u> for styrene, phenol, and a number of fibers and plastics.

A. 开始材料   B. 原始材料

C. 第一材料   D. 起始材料

4. However, there are many chemicals <u>derived from</u> these same aromatics; thus many aromatic petrochemicals have their beginning by selective extraction from naphtha or gas – oil reformate.

A. 产生   B. 来自   C. 衍生   D. 派生

5. Carbon black, also classed as an inorganic petrochemical, is made predominantly by the <u>partial</u> combustion of carbonaceous (organic) material in a limited supply of air.

A. 部分的   B. 偏差的

C. 不完全的   D. 偏离的

IV. Put the following sentences into Chinese.

1. A petrochemical is any chemical (as distinct from fuels and petroleum products) manufactured from petroleum (and natural gas) and used for a varie-

ty of commercial purposes.

2. Primary petrochemicals include olefins (ethylene, propylene and buta-diene) aromatics (benzene, toluene, and the isomers of xylene); and metha-nol.

3. When aromatics are needed for petrochemical manufacture, they are ex-tracted from the reformer's product using solvents, such as glycols (eg, the Udex process) and sulfolane.

4. Of the inorganic chemicals, ammonia is by far the most common.

5. Ammonia is produced by the direct reaction of hydrogen with nitrogen; air is the source of nitrogen.

V. Put the following paragraphs into Chinese.

1. Petrochemicals are generally divided into three groups: (1) aliphatics, such as butane and butene; (2) cycloaliphatics, such as cyclohexane, cyclo-hexane derivatives, and aromatics (eg, benzene, toluene, xylene, and naph-thalene); and (3) inorganics, such as sulfur, ammonia, ammonium sulfate, ammonium nitrate, and nitric acid.

2. Hydrogen sulfide is a constituent of natural gas and also of the majority of refinery gas streams, especially those off – gases from hydrodesulfurization processes. A majority of the sulfur is converted to sulfuric acid for the manufac-ture of fertilizers and other chemicals. Other uses for sulfur include the produc-tion of carbon disulfide, refined sulfur, and pulp and paper industry chemicals.

5.4 Synthetic Organic Chemicals

Guidance to Reading

Many substances are being used in place of certain naturally occurring ma-terials because either the natural product is unobtainable in sufficient quantity or because the physical properties of the synthetic substance have been chosen in order to provide the maximum degree of usefulness in the field chosen for their application. In many cases, the synthetic product is the result of a deliber-ate attempt to imitate some rare natural material. Moreover, it is intended to show that the internal molecular structure of the synthetic products shows them with the very properties which are their characteristic attributes. In this way, these substances have been built up so as to possess all the advantageous prop-

erties of the naturally occurring material without those disadvantages so gener-
ally characteristic of natural raw materials.

卟 Text

The global synthetic organic chemicals industry produces a continual stream of innovative products and process technologies. Through technological interlink – age, these products and processes percolate throughout the industrial economy, touching upon and transforming a myriad of sectors, even those apparently far removed from chemical manufacture.

The synthetic organic chemicals industry supplies materials to a wide range of industrial operations. These include plastics, fibers, and solvents, biochemical agents, **food chemicals**, and materials for construction. Those sectors increasingly affected by organic chemicals include the fastest growing and pivotal industries operating in the early twenty – first century: **electronics and telecommunications, the automotive and transportation industry**, biotechnology, agriculture, **environmental remediation**, and the food sciences.

The Earlier Development of Synthetic Organic Chemicals

The first organic chemicals were naturally occurring materials derived either from animals or plant by products. These included dyes and textiles, herb – based medicinals, **naval stores** and solvents, **cellulose – based products** including paper goods, and a variety of **resins** and **coatings** for commercial applications. The beginning of the modern synthetic organic chemical industry can be traced to the middle of the nineteenth century with the laboratory synthesis of the first dyes by William Perkin in England.

By the 1860s organic chemicals derived from coal – tar distillates had become an important branch of the chemical industry. Coal tar residues, when carefully distilled, supply basic aromatic building blocks for organic synthesis: benzene, toluene, and the xylenes. These materials were used as **intermediates** in the synthesis of dyes, medicinals, resins, and solvents. The development of improved ovens by the **metallurgical industry** for capturing increased volumes of coal tar distillates during the coking process helped spur the growth of the synthetic organic chemical industry. In the last quarter of the nineteenth century, the iron and steel industry was selling increasing volumes of these byproducts to local organic chemical producers.

At about this time, and continuing into the first part of the twentieth century, Germany took over from England as the world's dominant coal – tar – based organic chemical producing country. Germany obtained its raw materials mainly from its growing iron and steel industry. Also important to the future growth of the industry was Germany's ability to manufacture large volumes of high purity sulfuric acid, which was needed convert the coal tar intermediates into final products. Therefore, the emergence in the 1890s of the catalyst – based contact process for making high grade sulfuric acid, an innovation of the German firm Badische, proved a particularly critical development for the country's organics sector.

By the 1890s, the German chemical industry controlled nearly three – quarters of the world's coal tar chemicals markets. The major organic companies included Bayer, Badische, and Hoechst. In the 1920s, these and other of Germany's leading chemical producers would eventually combine into what would become the world's largest organic chemical company, I. G. Farbenindustrie. By the start of World War I, the German chemical industry supplied approximately 95% of total world output of synthetic organics. In 1914, the coal – tar – based dyestuffs alone that German exported to the United States are more than two and one half times the value of U. S. dye production.

Oil and Organic Chemicals

Coal tar distillates remained the dominant source of organics in Europe until the post World War II period. Even in the United States, coal tar held sway until the 1950s in the organics production of some major chemical producers, most notably, DuPont.

However, the prevailing trend for U. S. organics production after 1914 was the substitution of petroleum and natural gas for coal tars as the basis for synthetic organic chemical manufacture. This in turn resulted in the rising importance of U. S. organics vis – a – vis Germany.

The United States has remained the dominant oil production region for most of the twentieth century. This production has been critical to the establishment and growth of the U. S. organic petrochemical industry over the last six decades.

Between the 1860s and the mid – 1880s within the United States, most of the petroleum crude came from Pennsylvania, and more particularly, the Pittsburgh area. After 1885, oil production moved into the Midwest, and specifically Ohio and Indiana. Following World War I, the industry began to exploit the mid – continental fields of Oklahoma and Louisiana as well as the fields in New Mexico, Colorado, Wyoming, and California.

The fastest growing area for production of oil was the **Gulf of Mexico** region: between 1920 and 1941, its share of U. S. oil production had climbed from 6% to 16%. By the 1960s, Texas became the largest single producing state followed by California, Oklahoma, and Louisiana. By the 1980s, the Gulf region controlled 80% of U. S. oil production.

Structurally, oil is made up of hydrocarbon molecules containing various combinations of carbon and hydrogen. The configurations of these molecules depend on the number and arrangement of the carbon atoms. These may be linked in straight chains, branched chains, and circular, or ring formation. With respect to molecular weight, the lighter molecules (fewer carbon atoms) tend to be gases; the medium weight molecules are liquids; and the heavier (many carbon atoms) are heavy liquids and solids.

Hydrocarbons are divided into a variety of categories. Each category possesses a distinct molecular profile and, in turn, set of chemical and physical **properties**. Each class of hydrocarbons, therefore, has historically served different markets. Crude petroleum is composed of four major hydrocarbon groups: paraffins, olefins, naphthenes, and the aromatics.

In addition to the above hydrocarbon groups, crude oil also contains a number of other elements including sulfur, nitrogen, oxygen, metals, and **mineral salts**. Not all crudes are the same. They differ according to the geographical locations in which they are found.

The U. S. petrochemical industry specialized in the industrial use of the olefins or aliphatics. The first and to this day most important organic intermediate was the olefin, ethylene, first obtained commercially by Union Carbide (Carbide and Carbon Chemicals Corporation) in the 1920s. The mass production and bulk utilization of ethylene was followed by the use of propylene and butenes. The butenes, for example, were converted to **ketones** and their derivatives, and applied for a variety of applications as solvents in the automotive and other industries.

These olefins were the basic building blocks of the U. S. petrochemical industry. Between 1926 and 1939, the number of industrial compounds derived from ethylene increased from 5 to 41. During the same period the number of compounds derived from propylene grew almost tenfold, from 7 to 68.

By the 1940s and 1950s, U. S. petrochemicals moved into other chemical groups, including the diolefins, acetylene, paraffins, and finally the aromatics, which competed successfully against the coal – tar based materials. In the period up to 1950, petrochemical development was almost exclusively based in the United States. In the 1950s, petrochemicals production began to spread into Europe and Asia based on U. S. technology design. Since 1960, the petrochemicals industry has become global.

The emergence of petroleum refining during the twentieth century is closely linked with the growth and eventual world influence of the U. S. organic chemicals industry.

Words and Expressions

resins	n. 树脂
coating	n. 涂料
intermediate	n. 中间体,中间产品
ketones	n. 酮
property	n. 性质,属性

Technical Terms

food chemicals	食用化学品
electronics and telecommunication	电子通信
the automotive and transportation industry	汽车运输业
environmental remediation	环境治理
naval stores	松脂制品
cellulose – based product	纤维产品
metallurgical industry	冶金业
mineral salts	矿物盐
Gulf of Mexico	墨西哥湾

Language Focus

1. Coal tar residues, when carefully distilled, supply basic aromatic build-

ing blocks for organic synthesis: benzene, toluene, and the xylenes.

"When..."在此处作时间状语,为省略结构,省略了主语和谓语"coal tar residues are",英语中当从句的主语和主句主语一致,且谓语动词为be动词时,可以省略。

2. Coal tar distillates remained the dominant source of organics in Europe until the post World War II period.

上述句子中含有"until",表示时间的先后顺序。"until"引导的分句表示主句的情况先于或延续到这一分句所表示的时间,主句所表示的情况或动作必定是持续性的。如果主句中的动词是持续性的,则可使用肯定与否定两种形式,如果主句中的动词是非延续性的,则只能使用否定形式。

3. Even in the United States, coal tar held sway until the 1950s in the organics production of some major chemical producers, most notably, DuPont.

[参考译文为:即使在美国,在20世纪50年代之前,煤焦油一直主导一些主要化工产品生产商的生产,最著名的是杜邦公司(DuPont)]。

本句中,词组"hold sway (over sb/sth)"意为"对……有很大的支配或影响力,占统治地位",表明煤焦油稳固的统治地位。

4. However, the prevailing trend for U. S. organics production after 1914 was the substitution of petroleum and natural gas for coal tars as the basis for synthetic organic chemical manufacture.

(参考译文为:然而在1914年以后,美国有机产品的主要发展趋势发生了转变,石油和天然气取代煤焦油,成为有机合成化工产品的基本原料。)

本句的基本句子结构为"the prevailing trend was...",其中"the substitute of...for..."结构作表语。

5. Each category possesses a distinct molecular profile and, in turn, set of chemical and physical properties.

(参考译文为:"碳氢化合物分为很多种类,每一类都拥有独特的分子结构和相应的化学与物理性质"。)

句中,"each"和"every"都有"每一个"之意,但两者在使用上有一定区别。"each"突出个体独立性,"every"多指具有一定普遍性的个体。

Reinforced Learning

I. Answer the following questions for a comprehension of the text.

1. Which industrial operations can the synthetic organic chemicals industry supply materials?

2. What could we get from coal tar residues?

3. What were the reasons for development of organic chemical in Germany?

4. What changes happened in U. S. organic production after 1914?

5. Where did most of the petroleum crude come from between the 1860s and the mid – 1880s within the United States?

II. Multiple choice: choose the correct one from the alternative answers to give the exact meaning of the word underlined.

1. The global synthetic organic chemicals industry produces a continual stream of innovative products and process technologies.

  A. river   B. water   C. supply   D. chain

2. Through technological interlink – age, these products and processes percolate throughout the industrial economy, touching upon and transforming a myriad of sectors, even those apparently far removed from chemical manufacture.

  A. pass   B. move   C. penetrate   D. spread

3. Through technological interlink – age, these products and processes percolate throughout the industrial economy, touching upon and transforming a myriad of sectors, even those apparently far removed from chemical manufacture.

  A. separated   B. different   C. irrelevant   D. away

4. The first organic chemicals were naturally occurring materials derived either from animals or plant by products.

  A. happening   B. existing   C. affecting   D. evolving

5. The development of improved ovens by the metallurgical industry for capturing increased volumes of coal tar distillates during the coking process helped spur the growth of the synthetic organic chemical industry.

  A. catching   B. seizing   C. controlling   D. getting

6. However, the prevailing trend for U. S. organics production after 1914 was the substitution of petroleum and natural gas for coal tars as the basis for synthetic organic chemical manufacture.

  A. popular   B. widespread   C. influential   D. super

7. This in turn resulted in the rising importance of U. S. organics vis – a – vis Germany. The U. S. petrochemical industry specialized in the industrial use

of the olefins or aliphatics.

  A. majored B. centralized C. developed D. progressed

  8. The <u>configurations</u> of these molecules depend on the number and arrangement of the carbon atoms.

  A. shape B. outline C. layout D. design

  9. With respect to molecular weight, the lighter molecules (fewer carbon atoms) tend to be gases; the medium weight molecules are liquids; and the heavier (many carbon atoms) are <u>heavy</u> liquids and solids.

  A. large B. weighted C. solid D. stick

  10. Each category possesses a distinct molecular <u>profile</u> and, in turn, set of chemical and physical properties.

  A. state B. view C. impression D. structure

Ⅲ. Multiple choice：read the four suggested translations and choose the best answer.

  1. The beginning of the modern synthetic organic chemical industry can <u>be traced to</u> the middle of the nineteenth century with the laboratory synthesis of the first dyes by William Perkin in England.

  A. 追踪到 B. 追溯到 C. 调查到 D. 返回到

  2. At about this time, and continuing into the first part of the twentieth century, Germany <u>took over</u> from England as the world's dominant coal – tar – based organic chemical producing country.

  A. 遗传 B. 掠夺 C. 接手 D. 取代

  3. In the 1920s, these and other of Germany's leading chemical producers would eventually <u>combine into</u> what would become the world's largest organic chemical company, I. G. Farbenindustrie.

  A. 融入 B. 合并为 C. 组合为 D. 连接成

  4. The mass production and bulk utilization of ethylene <u>was followed by</u> the use of propylene and butenes.

  A. 被……跟随 B. 先于 C. 在……之后 D. 带动

  5. These olefins were the basic <u>building blocks</u> of the U. S. petrochemical industry.

  A. 砌块 B. 砖块 C. 积木 D. 基础材料

IV. Put the following sentences into Chinese.

1. The global synthetic organic chemicals industry produces a continual stream of innovative products and process technologies.

2. Between 1926 and 1939, the number of industrial compounds derived from ethylene increased from 5 to 41.

3. In the period up to 1950, petrochemical development was almost exclusively based in the United States.

4. Not all crudes are the same. They differ according to the geographical locations in which they are found.

5. The emergence of petroleum refining during the twentieth century is closely linked with the growth and eventual world influence of the U. S. organic chemicals industry.

V. Put the following paragraphs into Chinese.

1. The global synthetic organic chemicals industry produces a continual stream of innovative products and process technologies. Through technological interlink – age, these products and processes percolate throughout the industrial economy, touching upon and transforming a myriad of sectors, even those apparently far removed from chemical manufacture.

2. At about this time, and continuing into the first part of the twentieth century, Germany took over from England as the world's dominant coal – tar – based organic chemical producing country. Germany obtained its raw materials mainly from its growing iron and steel industry. Also important to the future growth of the industry was Germany's ability to manufacture large volumes of high purity sulfuric acid, which was needed convert the coal tar intermediates into final products.

Chapter 6   Oil Refining and Environmental Protection

6.1   Refining Contamination (1)

🔲 Guidance to Reading

The petroleum refining industry has a significant influence on the total pollution of the environment by industrial discharges and wastes. In the operation of petroleum refineries, the atmosphere is polluted with hydrogen sulfide, sulfur dioxide, nitrogen oxides, carbon monoxides, hydrocarbons and other toxic substances. The fresh water by refineries in product cooling is returned to the original source of water containing crude oil, petroleum products, and mineral salts as contaminants. The extent of air and water pollution depends on the particular processing technology and control measures that are employed, and also on the scale of processing.

🔲 Text

Environmental problems arise, however, in dealing with crude oil and its products because the petroleum industry — like hardly any other branch of industry — maintains exploration, transport, and refining installations for crude oil that are scattered over the entire globe. The distribution and consumption of burning and transportation fuels are even more scattered. Since the geographical location of crude oil fields does not **coincide with** that of the large consuming and refining regions, enormous transportation distances exist. The large amounts handled means that the industry needs to take precautions in all areas with regard to air emissions, water and soil pollution and, to a lesser degree, noise. The problems of introducing uniform standards are caused by legal standards which differ from country to country depending on the local concentrations of industrial areas with their corresponding environmental pollution and population density.

Because of the high cost of technical measures for environmental conservation, harmonization of legislation is absolutely necessary. As these costs amount to more than 25 % of the total processing costs in the refineries, equal competition between different countries must be guaranteed.

The environmental problems in crude oil refining, storage and loading, and in the application of the products are discussed below separately, because the problems in the various fields differ.

Manufacturing Emissions

Although closed, gas – tight systems are generally used in refinery units, emissions into air and water cannot be completely avoided even with careful handling during refining and storage of the crude oil and its products. This is due to the management and control of the process and the properties of the products concerned.

Hydrocarbons are discharged into the air because of their high vapor pressure and they appear in refinery wastewater **effluents** because of their water **solubility**, which is, however, small. The **carcinogenic** aromatic hydrocarbons are particularly dangerous.

Further attention must be paid to the sulfur and nitrogen compounds originating from the **heteroatomic** compounds in the crude oil, both because of their smell and toxicity and because of the air pollution which arises in the form of SO_2 and NO_x emissions during firing in process plants.

Hydrocarbon emissions can arise in production plants during normal operation (1) from leaking flanges in the pipework system; (2) at the seals of **valves**, pumps, and **compressors**; and (3) in the course of sampling. In the case of an accident, gases are led in closed systems to **flares**, collected as far as possible in gas recovery systems, compressed and returned to the process. The remainder of the gases is burnt in elevated or ground flares at efficiencies which can exceed 99%. Liquid products are collected in closed slop systems that are equipped with pressure reservoirs and tanks and later returned to the production **circuit**.

In new plants measures to reduce emissions are taken during their construction, whereas continuous **retrofitting** is necessary in existing plants. Examples are **flangeless** piping, low – emission stuffing boxes, and seals, such as duplicated slide ring packings. For intensely odorous, poisonous, and carcinogenic products, more effective measures should be adopted (ie. fixed motor pump and special extractor).

Most hydrocarbon emissions in processing occur in the storage areas, ie, tank farms for crude, feedstocks, intermediate, and final products. Pressure – vacuum relief valves and floating cover tanks are generally used to reduce emissions. More recent developments are emission – free tank farms, where several fixed roof tanks fitted with internal floating roofs breathe into a closed system at one common **gasometer**, which normally absorbs all changes in the tank level. In the rare event of unusually large changes in the system (large amounts flowing into or out of storage, solar irradiation, heavy rain), the surplus quantity is burnt harmlessly in an associated flare or, in the case of a pressure drop, the system is topped up with inert gas.

Considerable amounts of hydrocarbons are also emitted in the loading facilities, especially in the loading of gasolines. Here low – emission or emission – free loading for transport by ship, rail, and road has become largely accepted.

Various methods are used:

• Vapor recovery, where in a closed system the displaced gasoline vapors are either returned to the product tank or collected in a gasometer and used for the firing of process plants.

• Regenerative adsorption of the vapors on suitable adsorbents.

• Recovery of the products in liquid form after cooling or washing out the vapors.

Care must be taken to prevent the formation of explosive gasoline—air mixtures which can occur in the road tanker to be loaded and in the **adjacent** piping and equipment.

This can be achieved by (1) keeping the concentrations outside the explosive range, (2) short transportation paths and exclusion of **ignition** sources, and (3) extremely strict control of the oxygen contents. Dispensing of gasoline from the road tanker to the service station is also increasingly carried out with vapor recovery between the road tanker and the installed storage tank of the service station.

The measures described reduced the atmospheric emissions from processing, storage, and distribution of refinery products in the mid 1980s in Western Europe to less than 8% of the total man – made HC emissions. The portion from the refineries themselves was in turn only a quarter of this value.

Hydrocarbons in Wastewater

Hydrocarbon – containing wastewater is unavoidably obtained at various points in the refinery because of the water content of the crude oil itself and because steam is employed in various processing steps. The total amount of hydrocarbon – containing wastewater in a normal refinery is of the order of 60 ~ 100m³/h.

This wastewater must be removed from the process units after separation of the oil phase, and led via a closed system to wastewater purification. Rainwater from exposed plant areas and from tank yards, and possibly **contaminated** cooling water from leaks or accidents must be treated in the same way.

Treatment in the wastewater purification system is carried out stepwise by:

• Mechanical separation (sieves, filters, oil – water separators).

• Physicochemical purification (stripping, flocculation, flotation).

• Biological treatment.

Biological wastewater treatment of hydrocarbons in refinery wastewaters is normally problem – free. However, the incoming streams and corresponding buffer volume must be continuously monitored to detect pollution by sulfur or nitrogen compounds and by oxygen – containing components such as **phenols**.

In many countries there is an increasing legal requirement for covered water treatment plants to avoid odor and for total nitrogen removal for further protection of surface waters (rivers, lakes, etc.). The latter usually requires an additional purification stage with increased residence time. After the biological stage, the water is clean and can be discharged into the receiving water.

🔲 **Words and Expressions**

coincide with	与……一致
effluent	n. 废水;污物;水流
solubility	n. 可溶性;溶解度
carcinogenic	adj. 致癌的
heteroatomic	adj. 杂原子的 n. 异质原子
valve	n. 瓣膜;真空管;阀门
compressor	n. 压缩机
flare	n. 火炬 v. 闪光;爆发;张开

circuit	n. 巡回;环形;电路;回路 v. 巡回
flangeless	adj. 无凸缘的
gasometer	n. 气量计
retrofitting	n. 改装;翻新
adjacent	adj. 邻近的
ignition	n. 点火;[化]灼热
contaminated	adj. 受污染的
phenol	n. 酚;石碳酸

Language Focus

1. Environmental problems arise, however, in dealing with crude oil and its products because the petroleum industry — like hardly any other branch of industry — maintains exploration, transport, and refining installations for crude oil that are scattered over the entire globe.

本句中"arise"为不及物动词,意为"出现",例如:Accidents often arise from carelessness. 事故往往起因于粗心。另外,句中"...like hardly any other branch of industry..."意为"不像其他的行业","hardly"为副词表示否定。

2. Since the geographical location of crude oil fields does not coincide with that of the large consuming and refining regions, enormous transportation distances exist.

(参考译文为:油田和原油主要炼制与消费区不在一起,因而需要长距离的运输。)

本句中"coincide with"表示"与……一致",例如:My religious beliefs don't coincide with yours. 我的宗教信仰跟你的不一样。另外,句中"that"用作代词表示"the location"。

3. As these costs amount to more than 25% of the total processing costs in the refineries equal competition between different countries must be guaranteed.

(参考译文为:这一经费占整个炼制成本的1/4以上,因而必须保证国家之间的公平竞争。)

本句中"amount to"表示"总计,等于",例如:His debts amount to 5,000 pounds. 他的债务共达五千镑。"guarantee"一词表示"保证",例如:I guarantee that this will not happen again. 我保证此类事情不会再发生。

4. Care must be taken to prevent the formation of explosive gasoline—air mixtures which can occur in the road tanker to be loaded and in the adjacent

piping and equipment.

（参考译文为：在油罐车和临近的管路及装置中，必须防止产生爆炸性汽油—空气混合物。）

本句中"to be loaded"为动词不定式做定语，意为"要载重的"，常见诸如此类的用法还有 Do you have something to eat? 你有没有能吃的东西？另外，"care must be taken..."译为"必须防止……"在科技英语文章中常用被动语态表达科学客观的观点。

5. In many countries there is an increasing legal requirement for covered water treatment plants to avoid odor and for total nitrogen removal for further protection of surface waters (rivers, lakes etc.).

［参考译文为：在许多国家，为了除臭和完全除氮进而保护地表水（河流、湖泊等），对地下水处理厂的法律要求日益增多。］

本句中"increasing"意为"不断增加的"，例如：The increasing unemployment caused social unrest. 不断增加的失业引起了社会骚乱。

6. Rainwater from exposed plant areas and from tank yards, and possibly contaminated cooling water from leaks or accidents must be treated in the same way.

本句的主干为"rainwater and contaminated cooling water must be treated in the same way"，意为"雨水和冷却水也必须用同样的方法处理"。句中"from"引导的介宾短语作定语，用来解释雨水和冷却水的来源。另外，句中"exposed plant"意为"露天炼厂"。

Reinforced Learning

I. Answer the following questions for a comprehension of the text.

1. Why do environmental problems arise in dealing with crude oil and its products?

2. Why is harmonization of legislation absolutely necessary?

3. What are the problems with sulfur and nitrogen compounds originating from the heteroatomic compounds in the crude oil?

4. What measures should be taken to prevent the formation of explosive gasoline – air mixtures?

5. Please list some of the treatments for wastewater purification.

II. Multiple choice: choose the correct one from the alternative answers to give the exact meaning of the word underlined.

1. Since the geographical location of crude oil fields does not coincide with that of the large consuming and refining regions, enormous transportation distances exist.

    A. districts   B. religion   C. release   D. process

2. As these costs amount to more than 25% of the total processing costs in the refineries equal competition between different countries must be guaranteed.

    A. granted   B. ensured   C. entrusted   D. enlarged

3. Hydrocarbon emissions can arise in production plants . . .

    A. show   B. rise   C. occur   D. find

4. In new plants measures to reduce emissions are taken during their construction, whereas continuous retrofitting is necessary in existing plants.

    A. actions   B. treasures   C. methods   D. ways

5. More recent developments are emission – free tank farms, where several fixed roof tanks fitted with internal floating roofs breathe into a closed system at one common gasometer, which normally absorbs all changes in the tank level.

    A. generally   B. hopefully

    C. obviously   D. exceptionally

6. Hydrocarbon – containing wastewater is unavoidably obtained at various points in the refinery because of the water content of the crude oil itself and because steam is employed in various processing steps.

    A. apparently   B. inevitably

    C. generally   D. confidently

7. However, the incoming streams and corresponding buffer volume must be continuously monitored to detect pollution by sulfur or nitrogen compounds and by oxygen – containing components such as phenols.

    A. carefully   B. continually   C. occasionally   D. constantly

8. In many countries there is an increasing legal requirement for covered water treatment plants to avoid odor and for total nitrogen removal for further protection of surface waters (rivers, lakes etc.).

    A. fixed   B. growing   C. temporally   D. lasting

Ⅲ. Multiple choice:read the four suggested translations and choose the best answer.

1. Dispensing of gasoline from the road tanker to the service station is also increasingly carried out with vapor recovery between the road tanker and the installed storage tank of the service station.

　A. 安装的　　　B. 储存的　　　C. 特制的　　　D. 废弃的

2. More recent developments are emission – free tank farms, where several fixed roof tanks fitted with internal floating roofs breathe into a closed system at one common gasometer, which normally absorbs all changes in the tank level.

　A. 自由排放　　　　　　　　B. 零排放

　C. 无限排放　　　　　　　　D. 充分排放

3. The remainder of the gases is burnt in elevated or ground flares at efficiencies which can exceed 99%.

　A. 多余　　　　B. 大约　　　　C. 左右　　　　D. 超过

4. The problems of introducing uniform standards are caused by legal standards which differ from country to country depending on the local concentrations of industrial areas with their corresponding environmental pollution and population density.

　A. 一致的　　　B. 制式的　　　C. 统一的　　　D. 唯一的

5. Hydrocarbons are discharged into the air because of their high vapor pressure and they appear in refinery wastewater effluents because of their water solubility, which is, however, small.

　A. 提取　　　　B. 释放　　　　C. 游离　　　　D. 喷射

Ⅳ. Put the following sentences into Chinese.

1. The distribution and consumption of burning and transportation fuels are even more scattered.

2. Although closed, gas – tight systems are generally used in refinery units, emissions into air and water cannot be completely avoided even with careful handling during refining and storage of the crude oil and its products.

3. Further attention must be paid to the sulfur and nitrogen compounds originating from the heteroatomic compounds in the crude oil, both because of their smell and toxicity and because of the air pollution which arises in the form of SO_2 and NO_x emissions during firing in process plants.

4. Considerable amounts of hydrocarbons are also emitted in the loading facilities, especially in the loading of gasolines.

5. Hydrocarbon – containing wastewater is unavoidably obtained at various points in the refinery because of the water content of the crude oil itself and because steam is employed in various processing steps.

V. Put the following paragraphs into Chinese.

1. The environmental problems in crude oil refining, storage and loading, and in the application of the products are discussed below separately, because the problems in the various fields differ.

2. This wastewater must be removed from the process units after separation of the oil phase, and led via a closed system to wastewater purification. Rainwater from exposed plant areas and from tank yards, and possibly contaminated cooling water from leaks or accidents must be treated in the same way.

6.2　Refining Contamination (2)

🔲 Guidance to Reading

The growth rates in petroleum refining capacity will be higher in the future; hence, the measures to combat the discharge of pollutants, if these measures maintain at the present level, will be entirely inadequate to prevent further contamination of air, water and sound. Action is needed now in mapping out a set of measures that can be applied, with acceptable levels of capital investment, to bring about not only a relative but also an absolute reduction in the industrial discharges to the environment from refineries. These measures must include the introduction of basically new refining process, improved units and equipment and advanced methods for organizing production.

🔲 Text

Since hydrocarbons are **water – soluble** (even if only to a small degree), their **penetration** into the soil with possible contamination of the groundwater must be carefully avoided, whenever crude oil or any of its products are handled.

Transport from the oil terminal to the refinery is carried out almost **exclusively** in underground pipelines, which are also the safest means of transport. The choice of high – grade steels as construction material, good insulation, cathodic corrosion protection, and continuous monitoring for leaks, including visual monitoring from aircraft or ground **inspection**, ensure a high level of safety. On difficult terrain and in areas of extreme temperature fluctuations, additional measures must be taken (pipeline compensation, elevated piles, intermediate tanks etc.).

The location of the refinery must be carefully selected with regard to possible dangers to drinking water. According to new legislation, all HC – handling units must be erected so as to prevent the discharge of spilled product to the underground, to adjacent streets, or canals; the storage tanks are placed in collection spaces which are made **impermeable** to oil using clay layers, plastic tilts, or concrete lining; in case of a leak they must be capable of receiving the entire tank contents.

If hydrocarbon contamination occurs in the soil, the affected portion of soil must be removed to prevent **subsequent** pollution of groundwater. If small amounts have escaped, the contaminated soil is usually **combusted** in **incinerating** plants. With larger amounts—and particularly if large areas are polluted with chemical residues and dangerous **refuse**—the damage must be treated **in situ**. Depending on the nature of the soil and the corresponding migration of the oil, this can be done by pumping off and purifying the contaminated water, possibly by additional injection of fresh water in adjacent wells.

Degradation of the oil by **microorganisms** is becoming increasingly important. This can also be done in situ or by excavating the soil and treating it externally .

Loading and storage of the products outside the refinery are subject to similar regulations. However, these are still very different in individual countries. As a result of the large number of external distribution depots for transport fuels and heating oils and of service stations, and because of the enormous number of oil – heated households, special care must be taken against **overfilling** and escape of products due to leaks. **Corrosion – resistant** and nonageing steels, plastic – lined steel tanks, and novel, glass – fiber – reinforced, plastic tanks are widely used. These tanks must also be constructed such that the whole volume can be collected if leakage occurs (double – walled tanks).

In some countries with a population density and dense housing and industrial areas, the regulations which apply to pure hydrocarbons are also applied to the handling of process water with much lower HC – concentrations.

Sulfur and Nitrogen Compounds. As a natural product, crude oil also contains heteroatomic compounds containing sulfur, nitrogen, and oxygen in addition to hydrocarbons. Whereas the nitrogen and oxygen contents are in the **ppm** range and play only a secondary role in atmospheric emissions, the sulfur content of the crudes can be as high as several percent. The distribution over the individual refinery fractions varies, but the content increases with increasing molecular size.

Sulfur Compounds. Sulfur and its compounds are catalyst poisons and **adversely** affect atmospheric emissions. Hydrogen sulfide, mercaptans, and disulfides are odor nuisances, and sulfur dioxide is formed during the combustion of crude oil products. Therefore, sulfur and its compounds must be removed or their contents reduced.

The light refinery products, liquefied petroleum gas and gasoline, must be almost completely sulfur – free; for diesel fuels and light heating oils, a substantial sulfur reduction to 0. 1% ~ 0. 5% is required by recent legislation.

Serious problems exist with heavy fuel oil, which is used almost exclusively as fuel in large industrial **furnaces** and power stations and leads to considerable SO_2 emissions. Many countries have established a maximum sulfur content in fuels of 1% ~ 2% (wt). This value can be reached without additional treatment only with a few, low – sulfur crude oils, whose supplies are limited.

Sulfur is generally removed from distillates by hydrodesulfurization, whereby the chemically bound sulfur is converted to hydrogen sulfide. The H_2S is then removed in a gas **scrubber**, converted to elemental sulfur in the downstream Claus process, and supplied to the chemical industry as a raw material.

The intense smell of the H_2S – containing gases to be processed and the high toxicity of H_2S, even at high **dilution**, means that precautions must be taken when handling. All sulfur – processing plants must be completely gastight: in the hazard zones, instruments and alarm devices are installed which automatically shut down the plants in case of danger.

Sulfur recovery in refineries has increased greatly as the result of the lower permitted sulfur content of the products. In conventional Claus plants with two reactors connected in series the conversion of the H_2S feed is ca. 95 % , a third reactor increases the conversion to 96 %. For a further reduction of the H_2S content an after treatment step is required. Almost complete removal that is required in various countries (eg, Germany 10 ppm H_2S in off – gas) can only be achieved with an additional high – temperature combustion step.

Nitrogen Compounds. The nitrogen content of most crude oils is relatively low, and in the distillates it is reduced during hydroprocessing to a few ppm. The residual nitrogen content causes no further difficulty in the use of the products.

Ammonia, which is formed in the hydrogenation steps and added in many refineries for process control in various process stages (pH adjustment), enters the wastewater and can be removed in the biological stage of wastewater purification.

Noise. In the past, noise pollution in the neighborhood of petroleum refineries played only a secondary role. Worldwide, pumps and compressors were installed not in closed buildings but in the open air, because of the flammability of petroleum and its products. However, because the plants were smaller at that time, the noise emission of the units was relatively small. In addition, the sound radiation from furnaces and their burners produced relatively low sound levels in the refinery surroundings, because of the compact construction of the main process plants that were simultaneously screened by low noise auxiliary installations (tank farms) at the **periphery**.

Problems arose, though, in densely populated areas with insufficient distances between industrial and residential areas. Most noise problems were caused by flare noise in the event of process disturbances.

Owing to growing environmental awareness in the public and a correspondingly greater strictness of legislation, further measures will have to be taken in the future to reduce sound emissions. This is also necessary for technical reasons, because the complex conversion refinery maintains more and larger units with greater furnaces and higher – powered gas compressors with high – speed steam **turbines** etc.

Important measures for noise reduction could cover the following items:

● Low – noise burners and additional noise insulation on process heaters and piping.

● Sound hoods on the drive motors and turbines of pumps and **compres-**

sors. A complete "in – housing", however, may also give rise to safety problems, because fire fighting is more difficult and the danger of explosion is greater in closed rooms.

● Sound insulation on control valves.

● Low – noise flare stack tips for elevated flares or additional ground flares.

A noise reduction by 10 dB, ie, a reduction of the observable noise level by about one – half, is possible according to the states of the art, though only with considerable expenditure.

Words and Expressions

water – soluble	adj. 可溶于水的
penetration	n. 进入
exclusively	adv. 专门地;排他地
inspection	n. 视察;检查
impermeable	adj. 不能渗透的
subsequent	adj. 随后的;后来的
combust	v. 燃烧
incinerating	adj. 焚烧的
refuse	n. 废物,垃圾
in situ	adv. 在原地(就地,在现场)
microorganisms	n. 微生物
degradation	n. 降解
overfilling	n. 过量灌装
corrosion – resistant	adj. 耐蚀的
ppm	百万分之几
adversely	adv. 不利地,有害地
furnace	n. 炉
scrubber	n. 擦洗工具
dilution	n. 稀释
periphery	n. 周边;边缘;外围
turbine	n. 涡轮

Language Focus

1. Since hydrocarbons are water – soluble (even if only to a small de-

gree) , their penetration into the soil with possible contamination of the ground-water must be carefully avoided, whenever crude oil or any of its products are handled.

[参考译文为：由于烃类可溶于水(即使溶解度很小)，对原油及其产品进行加工处理时，必须小心避免烃类渗透到土壤中，污染地下水。]

句中"must be carefully avoided"和"are handled"为被动语态，在科技文章当中被动语态的使用可表示描述的客观性。

2. The location of the refinery must be carefully selected with regard to possible dangers to drinking water. According to new legislation, all HC - handling units must be erected so as to prevent the discharge of spilled product to the underground.

句中"with regard to"表示"关于，就……而言，考虑到……"，本句意为"由于炼油厂可能污染附近饮用水源，所以选址应该谨慎"。例如：I have nothing to say with regard to your complaints. 对于你的投诉，我无可奉告。

3. Loading and storage of the products outside the refinery are subject to similar regulations.

句中"be subject to"表示"受某种条件的制约"，例如：We are subject to the law of the land.

我们须遵守当地的法律。

4. Serious problems exist with heavy fuel oil, which is used almost exclusively as fuel in large industrial furnaces and power stations and leads to considerable SO_2 emissions.

(参考译文为：重质燃料油绝大多数用作大型工业炉燃料以及发电厂燃料，会排放大量 SO_2 ，存在严重问题。)

本句中"lead to"表示"导致"，科技英语中常用，例如：Such a mistake would perhaps lead to disastrous consequences. 这样一种错误可能导致灾难性的后果。

5. In the past, noise pollution in the neighborhood of petroleum refineries played only a secondary role.

本句中"noise pollution in the neighborhood of petroleum"意为"炼油厂对附近区域的噪声污染"。另外，"play a secondary role"，表示"处于次要地位"。

Reinforced Learning

Ⅰ. Answer the following questions for a comprehension of the text.

1. What would possibly occur if hydrocarbons penetrate into the soil?

2. What are the safest means of transport?

3. What should be carefully selected with regard to possible dangers to drinking water?

4. What standard is the recent legislation set for the light refinery products?

5. What are the important measures for noise reduction?

Ⅱ. Multiple choice: choose the correct one from the alternative answers to give the exact meaning of the word underlined.

1. Since hydrocarbons are water – soluble (even if only to a small degree), their penetration into the soil with possible contamination of the groundwater must be carefully avoided, whenever crude oil or any of its products are handled.

A. processed   B. handed   C. controlled   D. produced

2. Transport from the oil terminal to the refinery is carried out almost exclusively in underground pipelines, which are also the safest means of transport.

A. extremely   B. extrovertly   C. exceptionally   D. absolutely

3. On difficult terrain and in areas of extreme temperature fluctuations, additional measures must be taken (pipeline compensation, elevated piles, intermediate tanks etc.).

A. special   B. extra   C. specialized   D. preliminary

4. ... the storage tanks are placed in collection spaces which are made impermeable to oil using clay layers, plastic tilts, or concrete lining; in case of a leak they must be capable of receiving the entire tank contents.

A. essential   B. necessary   C. impervious   D. indispensable

5. Owing to growing environmental awareness in the public and a correspondingly greater strictness of legislation, further measures will have to be taken in the future to reduce sound emissions.

A. compatibly   B. respectively C. accordingly   D. similarly

6. These tanks must also be constructed such that the whole volume can be collected if leakage occurs (double – walled tanks).

A. construed   B. built   C. constrained   D. controlled

7. The distribution over the individual refinery fractions varies, but the content increases with increasing molecular size.

A. defines   B. differs   C. adapts   D. modifies

Chapter 6 Oil Refining and Environmental Protection

8. The light refinery products, liquefied petroleum gas and gasoline, must be almost completely sulfur – free; for diesel fuels and light heating oils, a <u>substantial</u> sulfur reduction to 0. 1% ~ 0. 5% is required by recent legislation.

    A. subtle   B. subsequent

    C. subordinate   D. a great amount of

9. This value can be reached without additional treatment only with a few, low – sulfur crude oils, whose supplies are <u>limited</u>.

    A. plenty   B. restricted   C. temporary   D. everlasting

10. Depending on the nature of the soil and the corresponding migration of the oil, this can be done by pumping off and purifying the contaminated water, possibly by additional injection of fresh water in <u>adjacent</u> wells.

    A. close   B. beside   C. neighboring   D. near

III. Multiple choice: read the four suggested translations and choose the best answer.

1. In addition, the sound radiation from furnaces and their burners produced relatively low sound levels in the refinery <u>surroundings</u>.

    A. 附近   B. 周围   C. 环境   D. 条件

2. Problems arose, though, in <u>densely populated areas</u> with insufficient distances between industrial and residential areas. Most noise problems were caused by flare noise in the event of process disturbances.

    A. 人口密度大的地区   B. 拥挤的地区

    C. 浓密的人口区   D. 浓厚人口的区域

3. In some countries with a population density and dense housing and industrial areas, the regulations which apply to pure hydrocarbons (i. e. , pro – ducts) are also applied to the <u>handling</u> of process water with much lower HC – concentrations.

    A. 处理   B. 掌控   C. 制造   D. 销售

4. Degradation of the oil by microorganisms is becoming increasingly important. This can also be done in situ or by <u>excavating</u> the soil and treating it externally.

    A. 勘探   B. 挖掘   C. 开采   D. 保护

5. A noise reduction by 10 dB, ie, a reduction of the observable noise level by about one – half, is possible according to the states of the art, though only with considerable <u>expenditure</u>.

A. 节省　　　　B. 消费　　　C. 开支　　　　D. 扩展

IV. Put the following sentences into Chinese.

1. As a result of the large number of external distribution depots for transport fuels and heating oils and of service stations, and because of the enormous number of oil – heated households, special care must be taken against overfilling and escape of products due to leaks.

2. The distribution over the individual refinery fractions varies, but the content increases with increasing molecular size.

3. This value can be reached without additional treatment only with a few, low – sulfur crude oils, whose supplies are limited.

4. The nitrogen content of most crude oils is relatively low, and in the distillates it is reduced during hydroprocessing to a few ppm.

5. A noise reduction by 10 dB, ie, a reduction of the observable noise level by about one – half, is possible according to the states of the art, though only with considerable expenditure.

V. Put the following paragraphs into Chinese.

1. Owing to growing environmental awareness in the public and a correspondingly greater strictness of legislation, further measures will have to be taken in the future to reduce sound emissions. This is also necessary for technical reasons, because the complex conversion refinery maintains more and larger units with greater furnaces and higher – powered gas compressors with high – speed steam turbines etc.

2. Ammonia, which is formed in the hydrogenation steps and added in many refineries for process control in various process stages（pH adjustment）, enters the wastewater and can be removed in the biological stage of wastewater purification.

6.3　Consuming Contamination

🔲 Guidance to Reading

The emissions from the combustion of refinery products outside the refinery are much greater than the emissions in the refining. Oil product transportation from production to the filling station brings about the HC emissions, which are *also caused by motor vehicles during refueling and running for high vapor pressure, and due to the unburnt hydrocarbons in the motor vehicle exhaust. CO, NO_x, unburnt hydrocarbons, and lead compounds are also emitted with the exhaust gas in addition to the combustion products of water and CO_2. The proportion in the emissions from the motor vehicle sector is one-third of the total output.*

Text

The emissions of SO_2 and NO_x from the combustion of refinery products outside the refinery are much greater than the emissions during refining itself. In addition, the HC emissions in the transportation field, both in **refueling** and in running motor vehicles, are also considerably greater than in the steps from production to the filling station.

Transportation Fuels

When gasoline is combusted in the **spark ignition engine**, varying amounts of CO, NO_x, unburnt hydrocarbons, and lead compounds are emitted with the exhaust gas in addition to the combustion products water and CO_2. The amounts differ according to the driving style, type of fuel, and engine construction. With the diesel engine, SO_2 and **soot** particles are also emitted. In addition to exhaust emissions, the emission of hydrocarbon vapor during refueling and running of motor vehicles is an important environmental pollution factor.

Motor Gasoline

Different factors cause emissions of hydrocarbons from motor vehicles: firstly, the unburnt hydrocarbons in the motor vehicle exhaust, and secondly, the hydrocarbons emitted during refueling and running, because of their high vapor pressure. The proportion in the emissions from the motor vehicle sector is one-third of the total output, corresponding to 12% of the total man-made emissions of organic substances.

Owing to the wide use of motor fuels, particular attention must be paid to benzene emissions. The limit for the benzene content of gasoline [1% ~ 5% (vol) is under discussion] presents the oil industry with problems, particularly in the production of unleaded gasoline, because the benzene content of various **blending stocks** can be considerably higher, depending on the quality or origin of the crude, up to 8% (vol) for high **severity reformate** and 18% ~ 40% (vol) for **pyrolysis** gasoline. Benzene extraction from HC-stocks does not solve this problem because of the cost and the question of reuse of the benzene recovered. Therefore, the evaporative losses must be minimized.

Diesel Fuel

The main exhaust gas problems of diesel vehicles concern the emissions of soot particles and SO_2. The NO_x and CO components are less important than in the spark ignition engine. Demands are increasingly being made for a reduction of these emissions, because of the smell of the exhaust gases, and because the soot particles may have a carcinogenic effect.

Burning Fuels

The largest proportion of refinery products worldwide is used as a fuel for domestic or industrial heating as well as for energy production. These are the fractions of LPG middle distillates and heavy fuel oil. Emissions of SO_2, NO_x, and CO that are associated with combustion, as well as emission of ash and unburnt carbon (in the form of soot) can be considerably influenced by selection and pretreatment of the products used and by the burner design. This is particularly true for the large heating installations operated with sulfur – rich heavy fuel oils.

Liquefied Petroleum Gas

The liquefied petroleum gases used as fuels are almost sulfur – free as a result of pretreatment in the refineries, and they burn without formation of soot. Their use presents no problem for the user.

Light Heating Oil (No. 2 Fuel)

Middle distillates contain up to 1. 5% (wt) sulfur, depending on the origin of the crude. Most of the sulfur can be removed in the refinery by hydrosulfurization.

The permissible sulfur limits for light heating oil in most countries are <0. 5% (wt), because of its use in domestic heating. In the EU, the sulfur content of domestic heating oil is now limited to 0. 2% (wt). If lower sulfur contents are to be achieved, desulfurization costs will rise steeply with the current range of crudes, according to studies by Concawe.

Heavy Fuel Oil (No. 6 Fuel)

The residues from crude oil refining are used as heavy fuel oil for industrial heating and energy production. Sulfur dioxide, nitrogen oxides, CO, and particulate emissions must be considered for heavy fuel oils; for environmental protection measures in this field.

Cost of Environmental Conservation

Because the understanding of environmental problems, technical solutions, and legislation are still in rapid development, the costs of environmental conservation have reached such a high level that they markedly influence the total processing costs. Besides the investment costs required for new plants (15% ~20% of the total cost is used for environmental conservation) and the conversion of old plants, there are continuous operating costs for energy, maintenance, personnel, etc. The most important measures in the processing field are:

● Gas washing systems and Claus plants for sulfur recovery from H_2S followed by fine purification.

● Closed systems for the discharge of gaseous (via a flare) and liquid hydrocarbons.

● Floating roofs and floating covers in storage tanks for crude oil and products.

● Hydrocarbon vapor recovery systems for the storage and loading of volatile products.

● Collection spaces for escaping hydrocarbons and **precipitation** water in production plants, **tank farms** and **loading plants**. **Drainage** to wastewater purification plants.

● Reduction of emissions (SO_2, NO_x) from the plant's firing installations.

● Closed sampling systems, laboratory analysis, and on – line instruments for pollutant measurement in air and wastewater.

The following measures are for the production of less polluting products:

● Hydrodesulfurization of gasoline and middle distillates.

● Reforming and **isomerization** for the production of high – octane components as a basis for unleaded gasoline.

● **Synthesis** of suitable components for unleaded gasoline (MTBE).

● Conversion of heavy residues into light, clean products.

● Separate storage and loading of unleaded types of transportation fuels.

🔲 Words and Expressions

refuel	v.	加油
soot	n.	煤烟

severity	n. 强度,苛刻度
reformate	n. 重整产品;重整油
pyrolysis	n. 裂解汽油
precipitation	n. 沉淀(作用)
drainage	n. 排水,污水
isomerization	n. 异构
synthesis	n. 合成

Technical Terms

spark ignition engine	火花点燃式发动机
blending stock	调和油料
tank farm	油库
loading plants	装载设备

Language Focus

1. When gasoline is combusted in the spark ignition engine, varying amounts of CO, NO_x, unburnt hydrocarbons, and lead compounds are emitted with the exhaust gas in addition to the combustion products water and CO_2.

此句中"when"引导时间状语从句,主句的主语为"varying amounts of CO, NO_x, unburnt hydrocarbons, and lead compounds"。句中"in addition (to)"表示"(除……之外)还;此外","in addition to A, B..."表示 A 和 B 性质或作用类似,属于同一范围。

2. Owing to the wide use of motor fuels, particular attention must be paid to benzene emissions.

(参考译文为:由于汽车燃料广泛应用,人们必须特别注意苯的排放。)

句中"owing to"表示原因,在科技英语中常用。

3. The limit for the benzene content of gasoline [1% ~ 5% (vol) is under discussion] presents the oil industry with problems, particularly in the production of unleaded gasoline, because the benzene content of various blending stocks can be considerably higher, depending on the quality or origin of the crude, up to 8% (vol) for high severity reformate and 18% ~ 40% (vol) for pyrolysis gasoline.

[参考译文为:汽油中苯含量应遵循怎样的标准(1% ~ 5%仍存在争议)仍是石油工业面临的问题,特别是在无铅汽油生产中,各种调和油料中苯含

量非常高,苯含量取决于原油的质量和来源,在高质量重整油中占8%,在裂解汽油中占到18%～40%。]

句中"present sb. with sth."表示"给……带来……,使……面临"。例如:The winner was presented with a trophy. 获胜者得到了奖品。"particularly"及其后面的介宾短语作状语,起补充说明作用。

4. Demands are increasingly being made for a reduction of these emissions, because of the smell of the exhaust gases, and because the soot particles may have a carcinogenic effect.

句中"demands"为此句主语,"are being made"是谓语,为被动语态的现在进行时,表示需求越来越多。

5. Besides the investment costs required for new plants (15% ～20% of the total cost is used for environmental conservation) and the conversion of old plants, there are continuous operating costs for energy, maintenance, personnel , etc.

[参考译文为:投资成本除了用于添置新设备(总成本的15%～20%用于环境保护)和改造旧设备外,还要支付运营过程中的能源、维修成本和人力费用等。]

本句的主体结构为"there be"句型,表达客观存在,在科技英语中常用。另外,"required for the new plant"为过去分词短语作后置定语,修饰"the investment costs"。

🔲 Reinforced Learning

I . Answer the following questions for a comprehension of the text.

1. What are the pollutants of gasoline combustion in the spark ignition engine?

2. What different factors cause emissions of hydrocarbon from motor vehicles?

3. Why does the use of liquefied petroleum present no problem for the user?

4. What are the permissible sulfur limits for light heating oil in most countries?

5. What are the reasons for increasing costs of environmental conservation?

II. Multiple choice:choose the correct one from the alternative answers to give the exact meaning of the word underlined.

1. In addition, the HC emissions in the transportation field, both in refuel and in running motor vehicles, are also considerably greater than in the steps from production to the filling station.

A. grades   B. processes   C. distances   D. ways

2. With the diesel engine, SO_2 and soot particles are also emitted.

A. given   B. sent   C. published   D. discharged

3. The proportion in the emissions from the motor vehicle sector is one – third of the total output, corresponding to 12% of the total man – made emissions of organic substances.

A. field   B. part   C. area   D. district

4. The limit for the benzene content of gasoline [1% ~5% (vol) is under discussion] presents the oil industry with problems, particularly in the production of unleaded gasoline, because the benzene content of various blending stocks can be considerably higher, depending on the quality or origin of the crude, up to 8% (vol) for high severity reformate and 18% ~ 40% (vol) for pyrolysis gasoline.

A. standard   B. degree   C. boundary   D. extent

5. The permissible sulfur limits for light heating oil in most countries are < 0.5% (wt), because of its use in domestic heating.

A. allowed   B. regulated   C. standard   D. usual

6. If lower sulfur contents are to be achieved, desulfurization costs will rise steeply with the current range of crudes, according to studies by Concawe.

A. series   B. scope   C. selection   D. grade

7. Because the understanding of environmental problems, technical solutions, and legislation are still in rapid development, the costs of environmental conservation have reached such a high level that they markedly influence the total processing costs.

A. obviously   B. significantly

C. importantly   D. increasingly

8. Besides the investment costs required for new plants (15% ~20% of the total cost is used for environmental conservation) and the conversion of old plants, there are continuous operating costs for energy , maintenance , person-

nel , etc.

    A. change   B. upgrade   C. rebuilding   D. change

9. Closed systems for the discharge of gaseous (via a flare) and liquid hydrocarbons.

    A. traced   B. enclosed

    C. importantly   D. continue

10. Collection spaces for escaping hydrocarbons and precipitation water in production plants, tank farms and loading plants.

    A. room   B. gap   C. volume   D. equipment

Ⅲ. Multiple choice: read the four suggested translations and choose the best answer.

1. Different factors cause emissions of hydrocarbons from motor vehicles: firstly, the unburnt hydrocarbons in the motor vehicle exhaust, and secondly, the hydrocarbons emitted during refueling and running, because of their high vapor pressure.

    A. 尾气   B. 废弃物   C. 排气管   D. 耗用物品

2. Demands are increasingly being made for a reduction of these emissions, because of the smell of the exhaust gases, and because the soot particles may have a carcinogenic effect.

    A. 泄漏的   B. 无力的   C. 废掉的   D. 耗尽的

3. The most important measures in the processing field are as follows.

    A. 设备   B. 方法   C. 方式   D. 途径

4. Gas washing systems and Claus plants for sulfur recovery from H_2S followed by fine purification.

    A. 在……之前   B. 被……跟随

    C. 随着   D. 顺着

5. Synthesis of suitable components for unleaded gasoline (MTBE).

    A. 合适的   B. 匹配的   C. 相应的   D. 搭配的

Ⅳ. Put the following sentences into Chinese.

1. In addition to exhaust emissions, the emission of hydrocarbon vapor during refueling and running of motor vehicles is an important environmental pollution factor.

2. The proportion in the emissions from the motor vehicle sector is one –

third of the total output, corresponding to 12% of the total man – made emissions of organic substances.

3. Owing to the wide use of motor fuels, particular attention must be paid to benzene emissions.

4. Demands are increasingly being made for a reduction of these emissions, because of the smell of the exhaust gases, and because the soot particles may have a carcinogenic effect.

5. The liquefied petroleum gases used as fuels are almost sulfur – free as a result of pretreatment in the refineries, and they burn without formation of soot.

V. Put the following paragraphs into Chinese.

1. Owing to the wide use of motor fuels, particular attention must be paid to benzene emissions. The limit for the benzene content of gasoline [1% ~5% (vol) is under discussion] presents the oil industry with problems, particularly in the production of unleaded gasoline, because the benzene content of various **blending stocks** can be considerably higher, depending on the quality or origin of the crude, up to 8% (vol) for high **severity reformate** and 18% ~ 40% (vol) for **pyrolysis** gasoline. Benzene extraction from HC – stocks does not solve this problem because of the cost and the question of reuse of the benzene **recovered**. Therefore, the evaporative losses must be minimized.

2. The largest proportion of refinery products worldwide is used as a fuel for domestic or industrial heating as well as for energy production. These are the fractions of LPG middle distillates and heavy fuel oil. Emissions of SO_2, NO_x, and CO that are associated with combustion, as well as emission of ash and unburnt carbon (in the form of soot) can be considerably influenced by selection and pretreatment of the products used and by the burner design. This is particularly true for the large heating installations operated with sulfur – rich heavy fuel oils.

6. 4　Chemical Product Contamination

🔲 Guidance to Reading

The inorganic sector was used for making large quantities of advanced explosives during World War I, whereas in World War Ⅱ new synthetic materials were employed for tires, parachutes, communications equipment, and the like. Since World War II *, synthetic rubber and fibers have been largely expanded in a number of ways. In the late* 1990s *, over* 90% *of all organics production was based on either petroleum or natural gas. After World War I , the problem of air pollution began to emerge , thus followed by various petrochemicals regulations. The petrochemical industry also understands that there is a direct correlation between environmental compliance and increased manufacturing productivity.*

Text

The Rise of Petrochemicals

The organic petrochemicals industry began in the east in the United States in the years following World War I. Over the following three decades, petroleum competed with coal and grain as feedstocks for the manufacture of synthetic organic chemicals .

In the 1920s and 1930s, most attention of the industry was focused on new process technologies involving the olefins: ethylene, butylene, and propylene. Petroleum refiners participated in organic chemical manufacture to only a limited extent prior to World War II. Most notably, in the early 1920s, Standard Oil of New Jersey (Exxon) developed a process for **synthesizing isopropyl** alcohol on a mass production basis. Additionally, in the 1930s, Shell produced synthetic ammonia from natural gas as well as selected organics using the off - gases of their refining operations as raw materials.

For the most part, early petrochemical manufacture was carried out by chemical producers proper. By the late 1920s, **Union Carbide**, through its division, Carbide and Carbon Chemicals Company, produced ethylene and its derivatives from natural gas found in West Virginia. Ethylene was obtained by the direct cracking of natural gas and liquid petroleum fractions. Propylene and the butenes were derived either as a byproduct of ethylene production or from refinery operations. Important synthetics developed at this time by Carbide included **ethylene glycol**, which was used as one of the first synthetic **antifreeze materials** for automotive **radiator systems**, as well as **ethylene oxide** , and the **vinyl plastics**. The **glycol ethers** were also produced and used in new surface coatings for automobiles.

World War II proved to be a watershed time for petroleum - based synthetic organics. Whereas World War I utilized the **inorganics** sector for making large quantities of advanced **explosives**, World War II consumed different sorts of materials for new applications. Tanks, trucks, and aircraft required volume production of high – octane fuel. Also consumed were new synthetic materials for tires, parachutes, communications equipment, **and the like**. It was during World War II that the petroleum refiners actively converted their petroleum processing capability to the making of strategically vital petrochemicals, most notably synthetic rubber and fibers.

In the post World War II period, petrochemicals expanded in a number of ways. Geographically, Europe and Asia increasingly adopted petroleum – based processing technology, much of it adapted from U. S. technology. In the United States as well, petrochemical production took root across a wider geographical area. By the late 1990s, important organic chemical centers were located in Texas, Louisiana, Oklahoma as well as Illinois, Ohio, Michigan and West Virginia.

In addition, more products, previously manufactured from coal or grain, began to be derived from the more economical petroleum – based processes . Moreover, established petrochemicals , such as ethylene and ammonia, were produced in larger and technically more advanced plants.

By the 1960s, over 70% of all synthetic organics manufactured globally was derived from petroleum or natural gas. By the 1970s, there were approximately 3,000 distinct chemicals commercially produced from fossil fuels. In the late 1990s, over 90% of all organics production was based on either petroleum or natural gas. These raw materials generated in excess of 6,000 specific commercial organic compounds. Through the 1990s, the United States has remained the world's main petrochemical center.

Catalysts and Petrochemicals

Catalysts have played a dominant role in synthetic organic chemicals manufacture since the early part of the twentieth century. Catalysts are generally composed of one or more metals and metal compounds. Their commercial function is to accelerate reactions to the point that they become economically feasible.

The early commercial catalysts, which were either iron or platinum – based, were developed in Germany, most notably for the mass manufacture of the inorganic **heavy chemical**, sulfuric acid. In the 1930s, the first catalytic cracking technology, introduced by the U. S. refiner, Sun Oil, went on line commercially followed in the next ten years by larger and more advanced catalytic cracking and reforming technologies. These designs culminated in the 1940s.

By the late 1940s, catalysts were used in the making of intermediates for such critical organic synthetics as advanced fuels, and in the polymerization production of vinyl plastics, nylon, synthetic rubber, and the **methacrylates**. In the late 1960s and early 1970s, catalysts entered the environmental arena as the central component in **catalytic converters** for the control of automotive emissions.

By the late 1990s, catalysts accounted for well over 60% of organic production and 90% of current organic chemical processes. In 1999, catalyst manufacture represented in excess of $\$ 10 \times 10^9$ in four large market segments: refining, polymerization, chemicals, and environmental remediation. In that year, more than 100 firms were engaged in the manufacture of catalysts internationally. More than half of these were based in the United States.

Environmental Trends

Environmental concerns first emerged in the petrochemicals industry at the end of World War Ⅱ. In 1945, the petrochemicals industry began recovering sulfur from the hydrogen sulfide present in natural and refinery gas. This was done because of the problem of air pollution and world sulfur shortages in the late 1940s and early 1950s. This recovery technology was based on novel and more efficient **absorbents**.

The manufacture of chemicals from petroleum is based on the ready response of the various compounds types to basic chemical reactions, such as oxidation, halogenation, nitration, dehydrogenation, polymerization, and alkylation. The release of chemical contaminants from petrochemicals to the environment is unavoidable.

The objective of the **Toxics Substances Control Act (TSCA)** is to allow **EPA (U. S. Environmental Protection Agency)** to regulate new commercial chemicals before they enter the market, to regulate existing chemicals when they pose an unreasonable risk to health or to the environment, and to regulate their distribution and use. By collecting data on chemicals in order to evaluate, assess, mitigate, and control risks that may be posed by their manufacture, processing, and use, EPA can ban manufacture or distribution in commerce use, require labeling, or place other restrictions on chemicals that pose unreasonable risks.

Important groups of organic chemicals that have come under regulatory control since the 1970s, and in particular with respect to clean air, include the **methyl – based** fuel **additives** and the **halogen hydrocarbons** in the form of the **chlorofluorocarbons**(CFCs).

In more recent years, organic chemical companies have been investing billions of dollars in environmental technology to meet rapidly growing regulatory requirements. Companies are going beyond this and attempting to integrate environmentally beneficial technologies into their processes and to incorporate greater degree of **recyclability** into their production designs. In part this is to avert potential future and costly environmental regulation. By these measures, they hope to convince the regulatory authorities that the industry can regulate itself.

There are also commercial considerations. The growing role of international standards through such bodies as the **International Standards Organization** (**ISO**) means that companies who are not environmentally compliant to the ISO 14,000 standards will be at a competitive disadvantage: customers will avoid doing business with them in favor of environmentally cleaner companies. The industry also understands that there is a direct correlation between environmental compliance and increased manufacturing productivity, a vitally important concern for organic chemical companies.

🔲 Words and Expressions

synthesize	v. 合成
isopropyl	n. 异丙醇
inorganic	a. 无机的
explosive	n. 炸药
and the like	诸如此类
methacrylate	n. 甲基丙烯酸酯
absorbent	n. 吸附剂, 吸收剂
recyclability	n. 可回收利用性, 再循环能力
methyl – based	n. 甲基类
additive	n. 添加剂
chlorofluorocarbons (CFCs)	n. 氯氟烃

🔲 Technical Terms

Union Carbide	联合碳化物公司

ethylene glycol	乙二醇
antifreeze materials	防冻剂
radiator system	散热系统
ethylene oxide	环氧乙烷
vinyl plastics	乙烯基塑料
glycol ethers	乙二醇醚
heavy chemical	重化学品
vinyl plastics	聚乙烯基塑料
catalytic converter	催化转换器
Toxics Substances Control Act（TSCA）	有毒物质控制法案
U. S. Environmental Protection Agency（EPA）	美国环境保护机构
International Standards Organization	（ISO）国际标准化组织
halogen hydrocarbons	卤代烃

Language Focus

1. Propylene and the butenes were derived either as a byproduct of ethylene production or from refinery operations.

（参考译文为：丙烯和丁烯则作为乙烯生产的副产品获取，或者从炼油过程中制得。）

本句中"derive"后接两个不同介词，表示不同意思，"derive as"意为"作为……而产生"，"derive from"则表示"从……产生"。

2. Also consumed were new synthetic materials for tires, parachutes, communications equipment, and the like.

本句为倒装句，正常语序为"New synthetic materials were also consumed for tires, parachutes, communication equipment, and the like."将动词"consumed"提前后，整个句子则要倒装，起到了一定强调作用。"and the like"意为"诸如此类的东西"。

3. By the late 1990s, catalysts accounted for well over 60% of organic production and 90% of current organic chemical processes.

（参考译文为：到20世纪90年代后期，催化剂在有机物生产中的比例大大超过60%，在目前有机化学品生产中占90%。）

句中"account for"表示"占……的比重"，例如：Petrochemicals today account for one fourth of all the chemicals made, in ten years this amount is expected to double. 石油化工产品现在已占所有化学制品的四分之一，十年后预计这个数目还要加倍。

4. Important groups of organic chemicals that have come under regulatory control since the 1970s, and in particular with respect to clean air, include the methyl – based fuel additives and the halogen hydrocarbons in the form of the chlorofluorocarbons (CFCs).

[参考译文为：从20世纪70年代开始，一些重要门类的有机化学品受到监管控制，特别是清洁空气方面尤为突出。受到监管的有机化学品包括甲基类燃料添加剂和以氯氟烃(CFCs)形式存在的卤烃。]

句中"that"引导定语从句修饰"important groups"。另外，"and in particular with respect to clean air"这一插入成分起了提示作用。

Reinforced Learning

I. Answer the following questions for a comprehension of the text.

1. How did Union Carbide produce ethylene, propylene, and the butenes in late 1920s?

2. What's the biggest difference between two World Wars in weapons?

3. How did catalysts play a role in environmental protection in the late 1960s and early 1970s?

4. What is the objective of the Toxics Substances Control Act?

5. Why did some chemical companies invest in environmental technology?

II. Multiple choice: choose the correct one from the alternative answers to give the exact meaning of the word underlined.

1. Over the following three decades, petroleum competed with coal and grain as feedstocks for the manufacture of synthetic organic chemicals.

A. during   B. upon   C. above   D. about

2. For the most part, early petrochemical manufacture was carried out by chemical producers proper.

A. very   B. completely   C. properly   D. particular

3. Geographically, Europe and Asia increasingly adopted petroleum – based processing technology, much of it adapted from U. S. technology.

A. changed   B. modified   C. rearranged   D. adopted

4. Through the action of promoters mixed industrial chemical transforma-

tions. Their commercial function is to accelerate reactions to the point that they become economically feasible.

    A. practicable   B. possible   C. reasonable   D. suitable

5. These designs culminated in the 1940s.

    A. peaked   B. produced   C. resulted   D. developed

6. Within the synthetic organic chemicals industry, a few large – scale production facilities account for the bulk of total production.

    A. size   B. shape   C. main part   D. body

7. Environmental concerns first emerged in the petrochemicals industry at the end of World War Ⅱ.

    A. appeared   B. grew   C. came   D. brought

8. Companies are going beyond this and attempting to integrate environmentally beneficial technologies into their processes and to incorporate greater degree of recyclability into their production designs.

    A. merge   B. mix   C. integrate   D. employ

9. In part this is to avert potential future and costly environmental regulation.

    A. turn   B. change   C. avoid   D. move

10. The growing role of international standards through such bodies as the International Standards Organization (ISO) means that companies who are not environmentally compliant to the ISO 14,000 standards will be at a competitive disadvantage: customers will avoid doing business with them in favor of environmentally cleaner companies.

    A. conform   B. tend   C. persuade   D. object

Ⅲ. Multiple choice: read the four suggested translations and choose the best answer.

1. The organic petrochemicals industry began in the east in the United States in the years following World War Ⅰ.

    A. 有机的   B. 组织的   C. 器官的   D. 机体的

2. Over the following three decades, petroleum competed with coal and grain as feedstocks for the manufacture of synthetic organic chemicals.

    A. 竞争   B. 对抗   C. 击败   D. 取代

3. Petroleum refiners participated in organic chemical manufacture to only a limited extent prior to World War Ⅱ.

A. 分担   B. 投资   C. 从事   D. 进入

4. In the United States as well, petrochemical production <u>took root</u> across a wider geographical area.

A. 扎根   B. 确立   C. 开始   D. 形成

5. In the 1930s, the first catalytic cracking technology, introduced by the U. S. refiner, Sun Oil, went <u>on line</u> commercially followed in the next ten years by larger and more advanced catalytic cracking and reforming technologies.

A. 在线   B. 按顺序   C. 投产   D. 操作

IV. Put the following sentences into Chinese.

1. For the most part, early petrochemical manufacture was carried out by chemical producers proper.

2. Whereas World War I utilized the inorganics sector for making large quantities of advanced explosives, World War II consumed different sorts of materials for new applications.

3. It was during World War II that the petroleum refiners actively converted their petroleum processing capability to the making of strategically vital petrochemicals, most notably synthetic rubber and fibers.

4. Geographically, Europe and Asia increasingly adopted petroleum – based processing technology, much of it adapted from U. S. technology.

5. In the late 1960s and early 1970s, catalysts entered the environmental arena as the central component in catalytic converters for the control of automotive emissions.

V. Put the following paragraphs into Chinese.

1. Important synthetics developed at this time by Carbide included ethylene glycol, which was used as one of the first synthetic antifreeze materials for automotive radiator systems, as well as ethylene oxide acetone, and the vinyl plastics.

2. The growing role of international standards through such bodies as the International Standards Organization (ISO) means that companies who are not environmentally compliant to the ISO 14,000 standards will be at a competitive disadvantage: customers will avoid doing business with them in favor of environmentally cleaner companies.

Chinese Translation and Key to the Exercises

第1章　美国炼油工业发展

1.1　早期石油

导语

工业时代之前,石油用处极其有限,几乎没有市场,是石油的照明时代开启了现代石油工业。洛克菲勒石油炼制行业的横向联合引发了行业竞争和垄断,最后标准石油公司被迫解体……

课文

工业时代之前,石油的用处极其有限。据史料记载,人类使用自然渗流的柏油沥青石油可追溯到公元前 3000 年古代美索不达米亚人时期,主要用于建筑泥胶、船舶防水及医疗膏药,这种油似乎也小有市场。后来,近东文化成就了天然原油蒸馏,获得了油灯燃料,也是"希腊之火"的基本原料,拜占庭帝国作为军事燃烧弹将其引进来抵抗进攻船只。在西方,原油渗流有限,主要供应医疗使用。同期,一个俄罗斯小型产业在有天然原油渗流的巴库发展起来,最初也用于医药。然而石油解决了因鲸鱼油危机和煤油稀缺而造成的 19 世纪灯油短缺问题,从而开启了现代石油工业。

石油的照明时代

19 世纪 50 年代,原油炼制实验表明,经过炼制的原油符合灯油的条件,问题是无法取得令人满意的产量。1859 年 8 月 27 日,德雷克的钻机在距宾夕法尼亚州泰特斯维尔镇不远油溪镇附近 69 英尺的地方开采出石油,于是石油时代诞生了。灯油市场的需求得不到满足,石油工业迅猛而显著地发展。原油产量在 1859 年才 2000 桶,10 年后猛增到 480 万桶;1871 年产量达到 535 万桶。在 19 世纪 60 年代到 1900 年之间,美国原油和石油产品出口从其生产总额的三分之一提高到四分之三。

该行业入行容易,资本投资最少。提炼技术十分原始,包括在蒸馏器中加热原油以此获得所需的煤油组分;其他组分,如汽油馏分通常被排放到溪流或地面沟渠里。宾夕法尼亚州法院对开采法的改革也导致油田产生浪费行为。几位不同的土地所有者或承租人可能会共享一个地下油田。好比打猎比赛中，如果将猎物引诱到另一个人的土地，那么那个人就可以合法捕获猎物。根据英国普通法的先例，法院支持产权人"捕获"流入自家土地的石油。由于当时对石油地质了解有限，使人们错误地认为石油像河水一样在地下流淌。每个业主或承租人都加紧开钻采油，以免邻居先把油抽尽。石油市场呈现出石油时而过剩时而稀缺的特点。19世纪后期，石油战线从宾夕法尼亚州、俄亥俄州和印第安纳州向西推进得克萨斯州和加利福尼亚州，到20世纪时到达了中部大陆，然而这种浪费石油的做法依旧继续着。

竞争和垄断

年轻的约翰·洛克菲勒是1863年进入这个混乱行业的。他在克利夫兰和别人合作食品批发业务并获得了经验，给俄亥俄州部队供应商品赚了不少钱。克利夫兰拥有几个区位优势，使其发展成重要的炼油中心。克利夫兰位于伊利湖，靠近宾夕法尼亚州的油田，拥有干线铁路，连接东部地区的主要市场，可以获得劳动力与金融资本。和今天一样，石油工业包含四大主要部分：生产、炼制、运输和销售。所有生产即将石油"从地下弄出来"，包括勘探、钻井、租赁、泵送、一次和二次采油技术及其他相关活动。炼制包括从天然原油提炼出可用产品的所有相关过程。最初炼制只包括分馏和加热原油分离组分，但今天已经发展到一系列复杂的专业步骤。运输是石油工业的第三要素，包括将原油从油田运到炼油厂，将成品油装船运往市场。在19世纪六七十年代石油行业发展初期，运输意味着以驳船、铁路列车或马车等方式输送石油。石油运输逐渐发展为以管道、公路油罐车及公海油轮等方式运输。销售包括把石油产品分销给消费者。

洛克菲勒早期主要投身于炼油行业，通过横向联合求发展，后期才采取垂直整合策略。开始是作为合伙人，后来在1870年洛克菲勒和合作方共同创建标准石油公司一起获胜，即后来称之为克利夫兰的征服。洛克菲勒运用精明甚至不道义的策略，成功操控了克利夫兰的炼油业。标准石油公司集中兵力于运输业，利用其强大的规模通过运用铁路回扣获得优先海运权，然后去努力通过一系列的卡特尔协议来消除竞争。洛克菲勒企图利用南方炼油公司乃至后来的国家炼油协会控制炼制业，成为19世纪后期企业经营策略的课本范例。当洛克菲勒发现非正式协议不足以令人满意地严格控制该行业时，他改变了方式，开始了并购。

到1878年，标准石油公司控制了美国炼油总产能90%以上。为了使该经济帝国步入正轨，洛克菲勒定于1882年同合伙人创建标准石油信托协，后来根据新泽西州法律于1899年重组为标准石油公司（新泽西州），合法控股公司。洛克菲勒在19世纪80年代实行反向整合,在俄亥俄州获得开采权,以自有管道进入运输业,并在国内销售自己品牌的煤油,同时也出口。近乎垄断的标准石油公司成为独立石油生产商最痛恨的敌人,他们发现自己的价格被洛克菲勒的购买力拉低了,小经营商不得不面对降价大战被迫出售石油,也拉低了独立石油商的价格。

虽然标准石油公司在州法院已遭受敌人和竞争对手多年的合法攻击,但1890年谢尔曼反托拉斯法的通过才使联邦挑战赢得根基。西奥多·罗斯福总统在位期间支持革新,政治气候发生变化,联邦调查接踵而至,最终标准石油公司于1911年解体为数个股份公司。洛克菲勒个人并没有受到太多伤害,因为他仍然是许多公司的股东。历史学家认为,从前的垄断只不过变成了大型纵向一体化公司的寡头垄断。

Key to Exercises

Ⅰ. 1. In a parallel development, a small Russian industry developed around the presence of natural seepage in the Baku region, initially also for medicinal use. However, it took a 19th – century shortage of lamp oil, precipitated by a whale oil crisis and relative scarcity of coal oil, to spark the modern petroleum industry.

2. Yes, it did.

3. Cleveland possessed several locational advantages that encouraged its development as an important refining center. Located on Lake Erie, close to the Pennsylvania oil fields, it had railroad trunkline connections to major eastern markets, a readily available workforce, and sources of financial capital.

4. The industry, then as today, consisted of four main functional sectors: production, refining, transportation, and marketing.

5. This did not hurt Rockefeller personally too much because he remained a stockholder in these many companies, and historians have argued that the former monopoly had simply transformed into an oligopoly of large vertically integrated companies.

Ⅱ. 1 – 5 CAABB 6 – 10 BDBDB

Ⅲ. 1 – 5 CBCBA

Ⅳ. 1. 问题是无法取得令人满意的产量。

2. 石油运输逐渐发展为以管道、公路油罐车及公海油轮等方式运输。

3. 该行业入行容易,资本投资最少。

4. 石油市场呈现出石油时而过剩时而稀缺的特点。

5. 洛克菲勒企图利用南方炼油公司乃至后来的国家炼油协会控制炼制业,成为19世纪后期企业经营策略的课本范例。

Ⅴ.1. 灯油市场的需求得不到满足,石油工业迅猛而显著发展。原油产量1859年才2000桶,十年后猛增到480万桶;1871年产量达到535万桶。在19世纪60年代到1900年之间,美国原油和石油产品出口从其生产总额的三分之一提高到四分之三。

2. 虽然标准石油公司在州法院已遭受敌人和竞争对手多年的合法攻击,但1890年谢尔曼反托拉斯法的通过才使联邦挑战赢得根基。西奥多·罗斯福总统在位期间支持革新,政治气候发生变化,联邦调查接踵而至,最终标准石油公司于1911年解体为数个股份公司。洛克菲勒个人并没有受到太多伤害,因为他仍然是许多公司的股东;历史学家认为从前的垄断只不过变成了大型纵向一体化公司的寡头垄断。

1.2 技术更新与能源过渡

🔲 导语

从顿钻钻井到旋转钻井,从轻裂解到热裂解,从一次采油到提高采收率技术,技术革新促进了石油生产。更多的高质量汽油取代煤油生产出来,以满足日益增多的汽车需求。对当时洛克菲勒的兼并策略以及替代性液体燃料你怎么看,是否必要?

🔲 课文

如果说1859年德雷克井的诞生标志着照明油工业的诞生,那么斯潘德尔托普油田则象征新能源时代的到来。斯潘德尔托普油田以及随后在西部地区发现的新油田为经济增长提供了丰富的石油,也使 Gulf、Texas 和 Sun Oil 等公司获得了巨大的成长空间。墨西哥湾地区未经开发,吸引了很多年轻公司,而得克萨斯州则反对标准石油垄断。斯潘德尔托普的沥青基原油生产出的煤油和润滑油质量不高,但生产的燃油还是令人满意的。在煤炭紧缺的西部地区,像 Gulf 和 Texaco 等这样的新公司很容易就将这种燃油从炼制、运输到销售集于一体。

1900年,标准石油公司(新泽西州)控制了86%的原油供应、82%的炼油产能以及在美国市场销售的85%的煤油和汽油。在1911年法院判令前夕,标准石油公司对原油生产的控制已经降到60%~65%,对炼油产能控制已降至64%。

此外,标准石油公司的各竞争对手目前向国内市场提供约70%的燃油,45%的润滑油,33%的汽油和蜡,以及25%的煤油。在斯潘德尔托普油田发现之后成立的新公司,如 Gulf 和 Texcao,连同 Sun 和 Pure 这些生机勃勃的老牌公司总共占据了硕大的市场份额。与此同时,曾在传统的煤油业务上投入巨资的标准石油公司逐渐把业务重心转移到汽油的生产和销售上。

几乎是与此同时,19世纪90年代,汽车作为一个新事物出现,开始显示其高端形象。1908年,在福特 T 型车的带领下,汽车批量生产,价格相对低廉,很快促进了石油行业的发展。1908年,美国汽车业的总产量为6.5万辆。随后不到10年的时间里,仅福特汽车公司每年售出的汽车量就高达50多万。在这一阶段,汽油产量和总价值都超过煤油。现在,石油行业越来越关注如何能提高汽油产量来满足需求。

当石油行业面临实际也或表面上的供应不足时,技术开始受到关注。早在20世纪初,旋转钻探等新技术的引进增加了原油产量,炼油技术革新使原油库存进一步扩大。然而由于汽油需求快速增长,20世纪初在得克萨斯州、加利福尼亚州和俄克拉荷马州新发现的这些石油资源远远满足不了需求。曾经丢弃的相对无用的较轻组分在每桶原油中通常占10%~15%。炼油厂通过保留更多重质煤油的方式增加汽油组分,但却生产出劣质的汽车燃料。

可以添加后来被称为高辛烷值产品来提高汽油的性能,这些高辛烷值产品来自于精选的高级原油和从高度饱和的天然气中生产出来的天然汽油。然而突破性进展是在1913年引进了热裂解技术时才取得的 。炼制业此前曾使用过光裂解法,即用加热法蒸馏将烃分子逐一进行重新排列,从19世纪60年代开始就是为了从原料中获取大量煤油。伯顿博士发现,通过大幅度提高温度和压力,分馏和裂化方法可能会比以前提高一倍的汽油产量。还有一个额外的好处是,这种汽油的质量普遍优于未热裂解的汽油。

丰富、匮乏和保护

从19世纪早期开始,不断有油田枯竭,继而又发现新油田,如此周而复始,因此石油工业一直面临周期性的油源丰富或油荒。1908年,美国地质调查局发布对未来美国石油供给的预测,并指出,如果按照目前的消耗水平,美国石油储量到1927年会枯竭。美国内政部长在1909年就向当时的塔夫脱总统表示,要及早应对即将到来的油荒问题。开采法的颁布促使人们急于开采石油,石油生产部门秩序混乱,石油开采也极难预测,因为井架林立的油田不断在向西部移动。在这种动荡的商业背景下,人们可以看到标准石油集团(新泽西州)的辩护律师从不同角度为该公司进行的反垄断辩护。他们坚持声称洛克菲勒的兼并战略(该行为被指责为恶性垄断)事实上是试图增加不稳定行业的稳定性和有序性。

20 世纪初,在西奥多·罗斯福政府的领导下,反垄断法登上政治舞台,开始发挥作用,新保存主义时代随之到来。保存方法(最好定义为对自然资源的明智利用)和与罗斯福同时代的塞拉俱乐部创始人约翰·缪尔提出保护主义者方法不同,区分开两者十分重要。罗斯福及其重要顾问并不想将美国的资源束之高阁,他们支持对资源进行积极规划并进行负责任的开发。人们应该理解美国地质调查局和矿务局(创建于 1910 年)在这种情况下的努力。这两个机构重视与石油相关的各种问题,但于 1914 年创建的矿务局石油部标志着联邦政府更加关注石油资源的保护。巴特尔斯维尔的俄克拉荷马石油实验站致力于研究油层动态、高效钻井作业、布井和二次采收技术,并通过颇具影响力的矿务局出版物来传播石油行业的相关知识。

20 世纪 20 年代,石油保存主义者开始鼓励油田的统一化管理或区域化管理,取代浪费开采法。这一管理模式鼓励每个油田作为合作单位参加,根据每个油田承租者的份额确定产油量。油田作业者减少钻井数量,控制采油速度,对石油伴生气或者利用或者重新注入产油层,而不是按照以往惯例将其燃烧或排入大气层。

20 世纪 20 年代初,石油储备即将枯竭,但需求量增加,促进了技术的发展。矿务局对提高采收率工艺和二次采油工艺进行了广泛调查,如在老油田中采用水驱法以提高产油量。为避免侵犯印第安纳州标准石油公司的波顿专利,各大石油公司各自开发了热裂解技术,增加了汽油产量。20 年代初也曾一度考虑过替换性液体燃料。标准石油公司(新泽西州)曾在 1922—1923 年间推出了一种含 25% 乙醇的混合汽油,随后以煤炭液化技术获得了德国基本专利权。在这一时期,西方国家还掀起了开发页岩油的热潮,留下了颇具趣味的奇特历史。不久,事实证明,所有这些替代性液体燃料是完全没有必要的,因为又一批新油田的发现削弱了人们对油荒的担忧。

🏭 Key to Exercises

Ⅰ. 1. If the Drake well in 1859 had trumpeted the birth of the illuminating oil industry, Spindletop marked the birth of its new age of energy. Spindletop and subsequent other new western fields provided vast oil for the growing economy but also enabled firms such as Gulf, Texas, and Sun Oil to gain a substantial foothold.

2. One could enrich the blend of gasoline with the addition of what would later be termed higher octane product obtained from selected premium crudes or

"natural" or "casing head" gasoline yielded from highly saturated natural gas.

3. The most important breakthrough, however, occurred with the introduction of thermal cracking technology in 1913. The industry had employed light cracking, the application of heat to distillation to literally rearrange hydrocarbon molecules, since the 1860s to obtain higher yields of kerosene from feed stock.

4. Roosevelt and his key advisers did not want to lock away America's resources but did favor planning and responsible exploitation.

5. In the 1920s, petroleum conservationists began to encourage unitization or the unit management of oil pools as a central alternative to the wasteful law of capture. This approach encouraged each pool to be operated as a cooperative unit, with the individual leaseholder's percentage share defining the amount of oil that could be pumped out.

Ⅱ. 1 – 5　CBACC　6 – 10　AACAD

Ⅲ. 1 – 5　BDBCC

Ⅳ. 1. 然而突破性进展是在 1913 年引进了热裂解技术时才取得的 。

2. 曾经丢弃的相对无用的较轻组分在每桶原油中通常占 10% ~15%。

3. 从 19 世纪早期开始,不断有油田枯竭,继而又发现新油田,如此周而复始,因此石油工业一直面临周期性的油源丰富或油荒。

4. 他们坚持声称洛克菲勒的合并战略(该行为被指责为恶性垄断)事实上是试图增加不稳定行业的稳定性和有序性。

5. 现在,石油行业越来越关注如何能提高汽油产量来满足需求。

Ⅴ. 1. 1900 年,标准石油公司(新泽西州)控制了 86% 的原油供应,82% 的炼油产能以及在美国市场销售的 85% 的煤油和汽油。在 1911 年法院判令前夕,标准石油公司对原油生产的控制已经降到 60% ~65%,对炼油产能控制已降至 64%。

2. 20 世纪 20 年代,石油保守主义者开始鼓励油田的统一化管理或区域化管理,取代浪费开采法。这一管理模式鼓励每个油田作为合作单位参加,根据每个油田承租者的份额确定产油量。油田作业者减少钻井数量,控制采油速度,对石油伴生气或者利用或者重新注入产油层,而不是按照以往惯例,将其燃烧或排入大气层。

1.3　美国石油炼制业

导语

石油炼制业是美国主要制造业之一,原油蒸馏主要由具有多功能炼油设备大型综合性公司进行。环保和安全法规对炼油行业产生的影响最大，迫使石油炼制工业在提高炼油工艺、减少污染物排放和改良产品组分上进行了大量投资。

🔲 **课文**

背景

从石油炼制产品在美国出口总值所占的比例来看,石油炼制业是美国领先制造业之一。虽然石油炼制业在美国经济中占有重要地位,但石油炼制公司和炼制机构数量却相对较少。当然这些数量随着信息来源的不同也有所不同。

1. 产品特性

石油炼制是用物理和化学方法先将原油蒸馏得到主要馏分,然后通过进一步分离转化成石油产品。炼油产品主要分为以下三大类:燃料油(汽油、柴油、燃料油、液化石油气、喷气燃料、残渣燃料油、煤油、焦炭);非燃料产品(溶剂、润滑油、润滑脂、石油蜡、凡士林、沥青、焦炭);化工原料(石脑油、乙烷、丙烷、丁烷、乙烯、丙烯、丁烯、丁二烯、苯、甲苯、二甲苯)。上述石油产品总量占美国能源总耗的40%(根据消耗的热量单位计算),并用作其他产品的主要原料,如用于化肥、农药、涂料、石蜡、稀释剂、溶剂、清洗液、清洁剂、制冷剂、抗冻结剂、树脂、密封剂、绝缘材料、乳胶、橡胶化合物、硬质塑料、塑料布、塑料泡沫和合成纤维。在美国90%的石油产品用作燃料,其中汽油又占43%。

2. 产业规模

一般来讲,石油炼制工业是由相对少量的大型设备构成。能源部报道中提到1994年有176处炼油厂,原油蒸馏能力约1500万桶/天。大型综合性公司具有多功能炼油设备,因此拥有美国大部分原油炼制能力。而处理能力低于5万桶/天的小型炼油厂在行业中也起着重要的作用,占有行业中50%的炼油设备,但此类炼油厂只拥有行业总原油蒸馏能力的14%。就其在经济方面的重要性而言,石油炼制业雇佣的人数相对较少。据普查局调查,1992年该行业直接雇用人数为7.5万,但也间接雇用了许多外部承包商,他们负责常规和非常规的炼制工作。据估计,1992年炼油厂出售了价值为1360亿美元的产品,约占美国整个制造业输出价值的4%。按炼油厂直接雇佣的人数来计算,该行业创造产品价值高达180万美元/人。相比之下,钢铁制造业工人当年创造产出价值的水平为24.5万美元。1992年统计局就业数据表明,60%的炼油厂有100名以上员工。

3. 经济趋势

美国是一个原油及其产品的净进口国。1994 年,美国使用的原油一半以上是进口的,石油产品的进口量约为 10%。预计原油进口份额还将增加,因为美国对石油产品的需求量增加,而国内原油产量却在减少。由于物流原因以及地区性产品匮乏,且供应商和炼油厂之间有长期的贸易关系,进口石油产品才在特定市场有了专门营业区。出口石油炼制产品约占美国炼制产品总额的 4%,主要包括石油焦、重质燃料油和轻质燃料油,出口原油约占美国生产原油和进口原油的 1%。

在过去 10 年里,美国石油炼制业承受了巨大的经济压力,这些压力来源于劳动力成本增加等大量因素以及新安全法和环保法规的颁布;而且 19 世纪 70 年代原油权益计划鼓励小型炼厂增强炼油能力,政府取消了原油补贴,增加了上述压力。1981 年年初,取消对原油定价和权益管制后,一切趋于合理化。在市场调节下,如果产能有剩余或利润空间渐小,低产能的工厂则会关闭。在诸多压力下,近几年许多低效工厂倒闭。据能源部调查,1982—1994 年间,美国 301 家炼厂降到了 176 家。大多数倒闭的企业都是原油处理能力小于 5 万桶/天的小型炼油厂,但也有一些较大型的炼厂因经济压力倒闭。

行业代表指出,越来越多的环保法规,特别是 1990 年颁布的《清洁空气法案》,是 20 世纪 90 年代对炼油行业产生影响最大的因素。尽管近年来关闭了一些炼油厂,但是炼制品产量一直保持相对稳定,在过去两年里还略有增加。而炼油厂产量增加是因为炼油产能的利用率提高,或在已有设备上提高炼油能力,而不是因为建立了新的炼油厂。

随着美国经济的增长,预计美国对石油炼制产品的需求也会缓慢增加,年均增幅约为 1.5%,低于预期的经济增长率。石油替代产品的开发,履行环保和安全法规成本的提高,导致石油产品价格上升,进而导致需求增长缓慢。

现在及未来环保和安全法规的调整将有望使石油炼制工业在改进某些炼油工艺、减少污染物排放和改良产品组分上进行大量投资。例如,为符合 1990 年清洁空气法案修正案对具体产品成分的规定,业内人士预计投资成本约 35 亿美元至 40 亿美元。因此有人担心一些炼油厂会部分或者全部关闭现有设备而非设备升级以满足新的标准,这样做或许更经济一些。实际上美国能源部和商务部预料到了整个 20 世纪 90 年代都会有炼油厂的关闭。但在现有设备合理使用、产能提高的情况下,总的原油蒸馏能力预计将保持相对稳定。炼制产品需求量的增加将由进口产品填补。

🔲 **Key to Exercises**

Ⅰ. 1. The primary products of the industry fall into three major categories: fuels (**motor gasoline**, diesel and **distillate fuel oil**, **liquefied petroleum gas** , jet fuel, **residual fuel oil** , kerosene, and coke) ; finished nonfuel products (solvents, **lubricating oils**, **greases**, **petroleum wax**, **petroleum jelly** , asphalt, and coke) ; and chemical industry feedstocks (**naphtha**, **ethane**, **propane**, **butane**, **ethylene**, **propylene**, **butylenes**, **butadiene**, **benzene**, **toluene**, **and xylene**) .

2. A relatively small number of people are employed by the petroleum refining industry in relation to its economic importance. The Bureau of the Census estimates that 75, 000 people were directly employed by the industry in 1992. However, the industry also indirectly employs a significant number of outside contractors for many refinery operations, both routine and non – routine. The value of product shipments sold by refining establishments was estimated to be $ 136 billion in 1992. This accounts for about 4 percent of the value of shipments for the entire U. S. manufacturing sector. Based on the number of people directly employed by refineries, the industry has a high value of shipments per employee of $ 1. 8 million.

3. The petroleum refining industry in the U. S. has felt considerable economic pressures in the past decade arising from a number of factors including: increased costs of labor; compliance with new safety and environmental regulations; and the **elimination** of government subsidies through the Crude Oil Entitlements Program which had encouraged smaller refineries to add capacity throughout the 1970s.

4. Increases in refinery outputs are attributable to higher utilization rates of refinery capacity, and to **incremental** additions to the refining capacity at existing facilities as opposed to construction of new refineries.

5. The rate of increase will average about 1. 5 percent per year, which is slower than the expected growth of the economy. This slower rate of increase of demand will be due to increasing prices of petroleum products as a result of the development of substitutes for petroleum products, and rising costs of compliance with environmental and safety requirements.

Ⅱ. 1 – 5 ADCAA 6 – 10 CAABD

Ⅲ. 1 – 5 ACCDD

Ⅳ. 1. 据估计,1992 年炼油厂出售了价值为 1360 亿美元的产品。

2. 大型综合性公司具有多功能炼油设备,因此拥有美国大部分原油炼制能力。

3. 1994 年,美国使用的原油一半以上是进口的,石油产品的进口量约为 10%。

4. 由于物流原因以及地区性产品匮乏,且供应商和炼油厂之间有着长期的贸易关系,进口石油产品才在特定市场有了专门营业区。

5. 1981 年初,取消对原油定价和权益管制后,一切趋于合理化。

Ⅴ. 1. 从石油炼制产品在美国出口总值所占的比例来看,石油炼制业是美国领先制造业之一。虽然石油炼制业在美国经济中占有重要地位,但石油炼制公司和炼制机构数量却相对较少。当然这些数量随着信息来源的不同也有所不同。

2. 随着美国经济的增长,预计美国对石油炼制产品的需求也会缓慢增加,年均增幅约为 1.5%,低于预期的经济增长率。石油替代产品的开发,履行环保和安全法规成本的提高,导致石油产品价格上升,进而导致需求增长缓慢。

1.4 炼制业的发展和石油产品开发

导语

石油炼制工业的发展及其产品的开发取决于市场需求,新的生活方式及对照明燃料、工厂燃料和汽车动力油日益增长的需求使石油消费量上升,同时也促进了石油炼制行业的发展和革新,开启了该行业的一体化发展。然而非常规石油炼制仍然是个极大的挑战。

课文

1. 炼制业的发展

石油在未炼制或天然状态下,就像许多工业原料一样几乎很少或者不能直接使用,其工业商品的价值只能在转化为销售产品后才能实现。即便如此,产品的类型仍然取决于市场的需求。因此,石油的价值与其产品的产量直接相关,同时也取决于市场的需求。

石油炼制,又称为石油加工,就是在高温高压下用蒸馏或化学反应的方法从原油当中获取能销售或有一定用途的成分和产品。合成原油是从焦油砂(油砂)沥青中提取的,在一些炼油厂中也作为原料使用。作为原油的基本成分,氢和碳是炼厂的主要原料成分,它们组合在一起可以形成上千种组分,并且因原油不同的质量以及炼厂不同的设备和加工方法,使原油呈现出不同的经济价值。

一般来说,原油一经炼制,将产生3种基本产品,这些产品分离成不同种类,但经常是交叉性馏分。蒸馏产生的馏分物数量取决于原油的原始成分和性质。气体和汽油组分形成沸点较低的产品,这些产品通常比沸点高的馏分物具有更高的价值,可以为化工业提供气体(液化石油气)、石脑油、航空燃料、发动机燃料以及化工原料。石脑油,即汽油和溶剂的前身,是从轻质和中质馏分油中提取的,也是石化工业的原料。中间馏分油是指处于石油沸点中间范围的产品,包括煤油、柴油燃料、馏分燃料油和轻瓦斯油。中间馏分油有时也包括含蜡馏分油和低沸点润滑油。原油的剩余物中还包括高沸点的润滑油、瓦斯油和渣油(原油的非挥发性成分)。渣油还能生产重质润滑油和石蜡,但更常用于生产沥青。目前谈到的石油复杂性体现在不同原油的轻质、中质和重质馏分的实际比例有很大不同。

常规石油通常是指可以从油气藏中抽吸出来的能够自由流动的液体,颜色有深有浅的原油。在一些油田里,油井中的压力充足,无需抽吸就能采收到石油。和常规气不同,重油流动性低,而且从地下储层中采收的难度更大。和常规石油相比,重油的黏度更高,API(美国石油学会)比重指数更低。此外,该类石油的一次开采经常需要热力开采。

重油的API比重指数通常小于20,含硫量按质量计算一般大于2%,超重油近乎呈固态,在周围环境中几乎无法自由流动。沥青焦油砂,常称为天然沥青,是超重油中的一种,通常是砂岩、石灰岩或泥质沉积物的裂缝和孔隙中的一种有机填充物,在这种情况下,有机质和伴生矿物基质称为岩沥青。渣油常缩写为"resid",是石油经过非破坏性蒸馏后去除所有挥发性物质之后的剩余物。

由于石油成分的热分解要高于350℃,所以蒸馏的温度通常低于345℃。减压蒸馏需要温度超过425℃,这个温度达到时就开始发生热分解,残留物通常称为沥青。从石油中提取的沥青往往跟天然沥青相似。蒸馏产生的沥青称为残留沥青。如果沥青是通过溶剂萃取的残留物或通过轻烃(丙烷)沉淀提炼而成的,也或者是通过吹制等其他方式制成的,那么该产品的名称就应该根据原料和制法而定,如丙烷沥青。高硫和低硫是指原油的近似硫含量,含硫量高的原油通常含有硫化氢(H_2S)和/或硫醇(RSH),这样的原油称为酸性原油。

2. 石油产品的开发

石油及其衍生物的使用是一种古老的工艺,如沥青和较重的非挥发性原油。事实上,石油的开发利用已有五千多年的历史记录。最早的使用记录是在美索不达米亚(古伊拉克),那时大家认识到这种非挥发性衍生物(沥青或天然沥青和人造沥青)可以防漏,也可用作珠宝黏合剂或建筑施工用的胶泥。大约2000年前,阿拉伯科学家发明了石油蒸馏法,这种方法随着阿拉伯人入侵西班牙而引入欧洲。当发现石油可用于照明以及烟煤的补充燃料时,石脑油开始受到关注,并越来越多地运用到战争中。希腊火药是一种石脑油—沥青的混合物,石脑油可以燃烧,沥青(或柏油)具有黏着性,可以延长燃烧时间。

现代炼制技术始于1859年,因为在宾夕法尼亚州发现了石油。当第一口井完工后,很快就有人租赁周边地区并开始大规模钻井。推动石油炼制工业发展的动力来自于生活方式的诸多改变,照明需求增加,工业革命中工厂动力燃料需求增加,汽车动力汽油需求增加,以及航空燃料的需求增加,这些都导致了石油用量的增加。同时,石油产品的结构也发生了变化。汽油和润滑油需求量的增加引起了人们对原油炼制的关注。因此原油炼制方法得以改变,炼油工业得到了创新和发展,最后促成了综合型石油炼厂的诞生。

对产品的不断需求,如液体燃料,是石油工业发展的主要推动力。一般来说,当产品是从石油中提取的一种成分,并且包含了大量的碳氢化合物分子时,这一成分就被归类为石油炼制产品。石油炼制产品包括汽油、柴油、取暖油、润滑油、蜡、沥青和焦炭等。相反,当产品仅含有一种或两种特定的高纯度碳氢化合物时,这一部分产品就被归类为石油化工产品。

某些特定的碳氢化合物在大气环境条件下以气体状态存在,如丙烷、丁烷、戊烷及其混合物,但在常温和适度压力下它们能转化为液体状态,称为LPG(液化石油气),这是一种石油炼制产品。

Key to Exercises

Ⅰ.1. Heavy oil differs from conventional petroleum in that its flow properties are reduced and it is much more difficult to recover from the subsurface reservoir. These materials have a much higher viscosity and lower API (American Petroleum Institute) gravity than conventional petroleum, and primary recovery of these petroleum types usually requires thermal stimulation of the reservoir.

2. Tar sand bitumen, often referred to as native asphalt, is a subclass of extra heavy oil and is frequently found as the organic filling in pores and crevices of sandstones, limestones, or argillaceous sediments, in which case the organic and associated mineral matrix is known as rock asphalt.

3. Tar sand bitumen, often referred to as native asphalt, is a subclass of extra heavy oil and is frequently found as the organic filling in pores and crevices of sandstones, limestones, or argillaceous sediments, in which case the organic and associated mineral matrix is known as rock asphalt. Asphalt, prepared from petroleum, often resembles native asphalt. When asphalt is produced by distillation, the product is called residual asphalt. However, if the asphalt is prepared by solvent extraction of residua or by light hydrocarbon (propane) precipitation, or if it is blown or otherwise treated, the name should be modified accordingly to qualify the product, eg, propane asphalt.

4. The increased needs for illuminants, for fuel to drive the factories of the industrial revolution, for gasoline to power the automobiles, as well as the demand for aviation fuel, all contributed to the increased use of petroleum. The product slate has also changed. The increased demand for gasoline and lubricants brought about an emphasis on refining crude oil. This, in turn, brought about changes in the way crude oil was refined and led to innovations and developments in the refining industry, thereby giving birth to the integrated petroleum refinery.

5. In general, when the product is a fraction that has been produced from petroleum and includes a large number of individual hydrocarbons, the fraction is classified as a refined product. Examples of refined products are gasoline, diesel fuel, heating oils, lubricants, waxes, asphalt, and coke. In contrast, when the product is limited to, perhaps, one or two specific hydrocarbons of high purity, the fraction is classified as a petrochemical product.

Ⅱ. 1 – 5 BACCC 6 – 10 CBAAC

Ⅲ. 1 – 5 CCAAA

Ⅳ. 1. 合成原油是从焦油砂(油砂)沥青中提取的,在一些炼油厂中也作为原料使用。

2. 一般来说,原油一经炼制,将产生 3 种基本产品,这些产品分离成不同种类,但经常是交叉性馏分。

3. 目前谈到的石油复杂性体现在不同原油的轻质、中质和重质馏分的实际比例有很大不同。

4. 在一些油田,油井中的压力充足,无需抽吸就能采收到石油。

5. 希腊火药是一种石脑油—沥青的混合物,石脑油可以燃烧,沥青(或柏油)具有黏着性,可以延长燃烧时间。

V. 1. 作为原油的基本组成成分,氢和碳是炼厂的主要原料成分,它们组合在一起可以形成上千种组分,并且原油不同的质量,以及炼厂不同的设备和加工方法,使原油呈现出不同的经济价值。

2. 某些特定的碳氢化合物在大气环境条件下以气体状态存在,例如丙烷、丁烷、戊烷及其混合物,但是在常温和适度压力下它们能转化为液体状态,称为 LPG——液化石油气,这是一种炼制产品。

第2章 石油炼制过程

2.1 石油简介

导语

石油是如何产生的？由什么组成？大量证据表明石油源于有机质，但石油的生成是非生物过程，由温度诱发，受有效时间影响。在石油储集过程中，运移发挥了重要作用，使富含有机质的生油岩和储油岩相互连通。油藏环境发生变化，石油成分也会随之发生变化。

课文

石油的来源

石油是一种自然生成的复杂混合物，主要成分是碳氢化合物，但通常也包含大量的氮、硫和氧，以及少量的镍、钒等其他元素。由于石油在地层下呈流体相态，因此是可以流动的，可能从很远的发源地汇集到一起，最终在可渗透的孔隙性储集岩中聚集。石油中含碳物质的生物起源广为人知，但未被普遍接受。但是石油包含多种所谓的化石或生物标记，这些物质拥有和生命系统相关的分子结构特性。此类化合物包括类异戊二烯、卟啉、甾烷、藿烷和很多其他烃类。石油中相对丰富的同类物常与生态系统中的该类物质相似。此外，混合物之间缺乏热动力平衡，石油和水环境下形成的沉积岩关系密切，表明石油是在低温（低于几百摄氏度）状态下产生的。石油的组成元素（碳、氢、氮、硫、氧）、同位素成分，以及近代沉积岩中出现的类石油物质都符合低温生成条件。支持石油生物起源的证据很多。

生物体产生种类繁多的有机混合物，包括大量生物聚合物，例如蛋白质、碳水化合物、木质素以及一系列低相对分子质量脂质。在生物体死亡后，全部或者部分有机质在水生环境中聚集，不同混合物在该环境中的稳定性大有不同，其中一些物质会在其他有机物（包括细菌）作用下在水体内新陈代谢，只有耐生化物质融入沉积物中。有机物质的存活量取决于很多因素，但主要是生态系统的自然氧化和还原作用。厌氧环境对于防止腐蚀十分有利，但是油气藏的形成需要的不仅仅是近期沉积物中产生的较低相对分子质量的烃类聚集。尽管有机物和沉积物中 $C_2 \sim C_{10}$ 碳氢化合物含量极低(约十亿分之一),但是它们在部分原油中所占比例高达50%或更多。

  石油的生成是非生物过程,由温度诱发,受有效时间影响。此过程为动力学一级反应,低温下每上升10℃左右,反应速率提升为2倍。干酪根生出的低相对分子质量混合物显示了近期沉积物的一个典型生物特性。因此,随着深度的增加,温度的上升,沥青失去其长链正构烷烃的奇偶优势和光学活性,同时失去了丁戊环烷烃的优势。这种趋势已在很多地区有完善记载。石油生成的过程可以基于第一能量守恒定律建立数量化模型来描述。

  大多数石油产于高渗透率和多孔储油岩中,此类岩石经过自然风化和风蚀作用后,去除了有机物等颗粒,渗透性和孔隙度大大提高。储油岩中的有机成分通常不足以产生满足商业需求的石油量。我们一贯认为石油产于富含有机质的生油岩中,然后部分沥青转移到储油岩中。这是生油岩的概念。显然,使富含有机质的生油岩和储油岩连通,运移发挥了关键作用。碳氢化合物在地下水中的溶解性太低,在石油运移过程中并不十分重要。近期的大多数研究强调了原油从生油岩中驱替出来。石油的生成及挤压过程中水的驱替作用导致生油岩内部石油高度饱和。石油持续遭受挤压,干酪根也不断转换为石油,于是油滴被迫离开生油岩挤入到邻近的渗透性运载岩中。岩石静压力高,能诱发近乎垂直的裂缝,为石油离开生油岩提供了重要途径。这种情况下产生的压力梯度能够克服浮力,石油可能会被迫向下挤出,从生油岩底部流出,当然也能从上部流出。生油岩中的烷烃相对不易吸收,因此最先排出。相比之下,氮硫氧混合物最易被岩石吸收,因此在排出的石油中这些混合物几乎没有了。石油运移效率差异较大,似乎取决于生油岩中有机质的含量。有机质含量高,运移效率就高。起决定性作用的是碳氢化合物(沥青)饱和程度,有机质含量低于大约1%(质量分数)的岩石无法产生足够的沥青,因此没有石油从这些岩石中运移出来,这些岩石不是有效生油岩。

  运载岩中的主要驱动力是浮力,石油会一直向上运移,直至进入构造圈闭中,或者在渗透率下降的地方,如地层圈闭中。运移距离可以高于100千米。石油在油藏首次聚集之后可能会重新运移。虽然最简单的情况就是重新换个位置,但如果油和气同时存在,可能会导致成分发生很大变化。当背斜油藏中油层上部有气顶,其石油多达溢出点时,从底部溢出的都是油,其相邻较浅的背斜只有油,就没有气顶了。这样不同的圈闭过程最终会导致油在较浅的圈闭中,而气在较深的圈闭中。

  环境发生变化,石油成分也会随之变化。石油埋藏越深,温度就越高,随之也会发生热熟化。大分子分解成分子碎片,按照这种趋势,石油密度越来越小,轻组分逐渐增多,由原油变为轻质油,然后变为湿气,最终变为干气。该过程中环化和芳构化反应同时进行,使氢含量越来越多。这些残余分子持续进行芳构化并增大,随着原油溶解性发生改变,在称为自然脱沥青的过程中,这些分子沉积在油藏中。固体沉积物中含很多镍和钒,所以可采原油的质量更高。

🔲 Key to Exercises

Ⅰ. 1. Petroleum is a naturally occurring complex mixture made up predominantly of carbon and hydrogen compounds, but also frequently containing significant amounts of nitrogen, sulfur, and oxygen together with smaller amounts of nickel, vanadium, and other elements.

2. After the death of the organism, all or part of this organic material may accumulate in aquatic environments where the various compounds have very different stabilities. Some are metabolized in the water column by other organisms (including bacteria) and only the biochemically resistant material is incorporated into sediments.

3. It follows essentially first – order kinetics where an increase of 10℃ roughly doubles the reaction rate at low temperatures. The low molecular weight compounds generated from the kerogen show none of the biological characteristics typical of compounds in more recent sediments. Therefore, at increasing depth, and hence increasing temperature, the bitumen fraction loses features such as odd – even predominance in long – chain normal alkanes, optical activity, and the predominance of four – and five – ringed cycloalkanes. These trends have been well – documented in many areas. The petroleum generation process can be treated quantitatively using models based on first – order kinetics.

4. Clearly, migration has a critical role in linking the organic – rich source rocks to the reservoir.

5. Buoyancy is the main driving force through the carrier beds and oils continue to move upward until stopped in a structural trap, or where permeability decreases as in a stratigraphic trap. Migration distances can be in excess of 100 km. Oil may be remobilized after its initial accumulation in the reservoir.

6. Thermal maturation of crude oil is brought about by the increasing temperature that accompanies increasing depth of burial. Some large molecules are broken down into smaller fragments and the trend is for an increasing percentage

of the lighter fractions as the oil progresses to lower densities in the sequence from oil, to lighter oil, to wet gas, and finally dry gas.

Ⅱ.1-5　ACBAA　6-10　DAABC

Ⅲ.1-5　ADAAC

Ⅳ.1. 石油中含碳物质的生物起源广为人知,但未被普遍接受。

2. 石油中相对丰富的同类物常与生态系统中的该类物质相似。

3. 生油岩中的烷烃相对不易吸收,因此最先排出。

4. 虽然最简单的情况就是重新换个位置,但如果油和气同时存在,则可能会使成分发生很大变化。

5. 石油埋藏越深,温度就越高,随之也会发生热熟化。

Ⅴ.1. 石油是一种自然生成的复杂混合物,主要成分是碳氢化合物,但通常也包含大量的氮、硫和氧,以及少量的镍、钒等其他元素。由于石油在地层下呈流体相态,因此是可以流动的,可能从很远的发源地地方汇集到一起,最终在可渗透的孔隙性储集岩中聚集。

2. 生物体产生种类繁多的有机混合物,包括大量生物聚合物,例如蛋白质、碳水化合物、木质素以及一系列低相对分子质量脂质。在生物体死亡后,全部或者部分有机质在水生环境中聚集,不同混合物在该环境中的稳定性大有不同。其中一些物质会在其他有机物(包括细菌)作用下在水体内新陈代谢,只有耐生化物质融入沉积物中。

2.2　石油的成分

导语

石油液态分子结构决定了原油性能。油品分类一直是基于沸点、密度、气味和黏度等物理性质。了解石油成分将有助于炼厂将原油转变为高价值产品,并且能够让环境学家考虑到石油暴露在环境中造成的生物影响,避免石油生产造成浪费和对环境的污染。

课文

石油包含三种形态:气态(天然气)、液态(原油)、固态或者半固态(沥青、焦油)。这一部分主要讨论石油液态分子结构对原油性质和表现的影响。原油在颜色、气味和流动性质上变化多端。这些性质通常反映了原油的起源。历史上油品分类一直是基于沸点、密度、气味和黏度等物理性质。根据相对密度大小来区分重质原油和轻质原油。轻质原油富含低沸点、含石蜡的碳氢化合物；重质原油含大量高沸点、类似沥青的分子。重油黏度更大，沸点更高，含有更多芳香族化合物，同时也包含大量的杂原子。同样可以用气味来区分低硫原油和高硫原油。

了解石油成分有助于炼厂将原油转变为高价值产品。起初，石油经过蒸馏后销售馏分，主要用于照明和润滑。原油以汽油、溶剂、柴油、航空燃油、取暖油、润滑油和沥青等形式出售，或者转化为石化原料出售，如乙烯、丙烯、丁烯、丁二烯和异戊二烯。现代炼制由复杂的加热、催化和加氢过程混合而成。转化过程包括焦化、加氢裂化和催化裂化，将大分子分解为小分子；加氢处理以减少杂原子和芳香烃，从而创造出有利于环境的产品；异构化和重整来重新排列分子成为高性价比产品，如高辛烷值汽油。

了解石油分子构成能够让环境学家考虑到石油暴露在环境中造成的生物影响。石油产于偏远地区，运输至距离市场较近的炼油厂。尽管只有少量的石油泄漏到环境中，却对环境造成潜在威胁。了解分子结构，不仅能确定污染源头，而且也有助于明白潜在有害物质造成的结果和影响。

原油有机化合物种类繁多且这些有机化合物分子大小分布范围很宽。种类如此繁多，哪怕对一种原油也无法完全用混合物—混合物的方式来表述。但是石油按照分子构成来描述可分为三类混合物：饱和烃、芳香烃和携带杂原子如硫、氧、氮的非烃化合物。每种类型里都包含几个族系。

分子类别

原油分子包含几种基本结构类型。因为原油包含 1~100 个碳原子并且可能会发生结合，同质异构类型统计数目会相当惊人。例如，C_{10} 链烃只有 75 种结构，而 C_{20} 异构体种类却超过 10^5。但是每个异构组的分布也只有几个结构起主导作用。

因为有环烷烃和芳香环，所以又增加了两个维度，使碳氢化合物异构体数量进一步增加。有人提出三维排列法，即分子描述可以基于芳香环的数量、环烷烃数量、烷基侧链含碳数等。从概念上而言，这相当于描述碳氢化合物三维分子架构。这也同时可能为杂环族化合物构建类似脉络。使用二维图像技术分析烷烃和芳香烃，可解释原油中的烷基替代物的变化。

大多数分析方法还无法分析整体原油分子特性。但是通过蒸馏法可以将原油分成相对原子质量较小的部分，以达到简化工作量的目的。1925 年创建的美国石油协会赞助使用这种创新方法，在一种原油馏分中就发现了成百上千种单体混合物。色谱技术的发展使得油品因其极性不同而实现精馏馏分分离。在美国石油学会的赞助下，美国矿务局将这种分离和测量技术扩展应用到（油品的）重组分分离。同时，人们可以从更高沸程的原油中定性和定量地分析各个组分。该技术综合了传统的空心柱吸收色谱图、胶化渗透色谱图、离子交换分离图等方法分离馏分。馏分中的化合物可通过质谱分析得出。

原油蒸馏所得各馏分量大有不同。常规原油中蒸馏出的中间馏分和减压瓦斯油的量基本相同。轻油中蒸馏出的石脑油量较多，而从重油中得到减压渣油量较多。石油馏分的典型分布与其沸点相对应。低沸点馏分主要由非极性饱和烃组成，其中的同分异构体也较少；高沸点馏分中则化合物种类较多，因此可能含有更多的同分异构体。随着沸点升高，形成了芳香环结构，首先是裸环，然后是越来越多的带有侧链和环烷环的芳香环。一般情况下含氧氮类的极性组分在低沸点馏分中很少，是其中的杂质；而在高沸点馏分中这种极性组分逐步增多。这一现象与石油馏分中硫氮分布和其沸点的关系相吻合。氮主要集中于高沸点馏分中，硫［不包括硫化氢（H_2S）和一些轻组分硫化物，如硫醇和石油气中的硫化物］的分布更广。氢碳之比没有随沸点升高而降低，表明随着沸点升高芳烃含量逐渐增加。金属、氧和氮元素主要出现在极性很强的高沸点馏分中。

分析方法

对各馏分的分子组成用不同方法进行分析。随着相对分子质量及复杂程度的增加，分析也需要更细化，包括对大多数种类石脑油的量化，对减压渣油平均相对分子质量的描述。对于石脑油，传统方法是依据其物理性质对每种化合物进行鉴别和区分。气相色谱法可以分别测量所有碳原子量少于8的化合物。当气相色谱法不能很好地对化合物进行定量分析时，可以将气相色谱法和质谱分析法联合使用。

Key to Exercises

I. 1. Crude oils may be called light or heavy in reference to relative density. Light crude oils are rich in low boiling and paraffinic hydrocarbons; heavy crude oils contain greater amounts of high boiling and asphalt – like molecules. The heavy oils tend to be more viscous, higher boiling, more aromatic, and contain larger amounts of heteroatoms. Likewise, odor is used to distinguish between sweet or low sulfur, and sour or high sulfur, crude oils.

2. The molecular composition of petroleum can, however, be described in terms of three classes of compounds: saturates, aromatics, and compounds bearing the heteroatoms sulfur, oxygen, or nitrogen.

3. A three – dimensional array in which the molecules could be described in

terms of the number of aromatic rings, the number of naphthenic rings, and the number of carbons in alkyl side chains has been proposed.

4. Techniques have been developed that use combinations of classical open – column adsorption chromatography, gel permeation chromatography, and ion – exchange separations to isolate fractions in which compounds could be identified by mass spectrometry.

5. Crude distillations yield different quantities in each fraction. About the same amounts are distilled into the middle distillate and vacuum gas oil from conventional crude oils. More naphtha is distilled from light crude oils and more vacuum residuum is obtained from heavy crude oils.

II.1 – 5　DDDDA　6 – 10　BDBBD

III.1 – 5　BCDAA

IV.1. 种类如此繁多,哪怕对一种原油也无法完全用混合物—混合物的方式来表述。

2. 因为有环烷烃和芳香环,所以又增加了两个维度,使碳氢化合物异构体数量进一步增加。

3. 同时,人们可以从更高沸程的原油中定性和定量地分析各个组分。

4. 常规原油中蒸馏出的中间馏分和减压瓦斯油的量基本相同。

5. 石油馏分的典型分布与其沸点相对应。

V.1. 了解石油分子构成能够让环境学家考虑到石油暴露在环境中造成的生物影响。石油产于偏远地区,运输至距离市场较近的炼油厂。尽管只有少量的石油泄漏到环境中,却对环境造成潜在威胁。了解分子结构不仅能确定污染源头,而且有助于明白潜在有害物质造成的结果和影响。

2. 原油分子包含几种基本结构类型。因为原油包含 1 ~ 100 的碳原子并且可能会发生结合,同质异构类型统计数目相当惊人。例如,C_{10} 链烃只有 75 种结构,而 C_{20} 异构体种类却超过 10^5 种。但每个异构组的分布也只有几个结构起主导作用。

2.3　石油炼制

🏭 导语

地质成因不同,原油的成分会有很大差异。石油炼制是一个复杂的过程,包括很多工艺过程,具体的工艺过程则取决于要炼制的原油和目标产品的性质。石油炼制一般包括两个阶段和一些辅助过程。

课文

　　石油由很多不同的烃类及少量杂质混合而成。地质成因不同,原油的成分会有很大的差异。石油炼制是一个复杂的过程,包括很多工艺过程,具体的工艺过程则取决于要炼制的原油和目标产品的性质。正因为这样,没有两家完全一样的炼油厂。部分产品从某些工艺过程中输出又返回,然后进入下一个工艺流程,又返回到上一个工艺,或与其他产品混合,从而得到成品。下面简要介绍石油炼厂中的主要生产过程。除了下面列出的以外,还有许多特殊工艺流程不能在这里一并提到,这些流程在控制污染物排放和满足产品特殊需求方面起着很重要的作用。

　　原油炼成石油产品可分为两个阶段和一些辅助过程。第一阶段是原油脱盐,蒸馏成不同组分。第二阶段由三种不同的下游工艺组成:合成、分裂与重塑。通过焦化、裂化、重整和烷基化等不同过程,下游工序将一些馏分转化成石油产品(重质燃料油、汽油、煤油等)。辅助工序可能包括污水处理、硫黄回收、添加剂的生产、热交换器清洗、排污系统、产品组合和储运等。

原油蒸馏和脱盐

　　炼油厂中最重要的一个工艺流程是先将原油蒸馏得到各种馏分。蒸馏包括加热、汽化、分馏、冷凝以及原料冷却。本文主要讨论常压蒸馏、减压蒸馏,相继使用这两个流程可以降低成本、提高效率。此外,本文还要讨论原油蒸馏前重要的第一步:脱盐处理。

原油的脱盐

　　原油分馏成各种馏分之前,通常先要脱掉腐蚀性的盐。脱盐过程也能除去一些导致催化剂失活的金属和固体悬浮物。脱盐过程包括将原油加热与水混合(原油为 3%~10%),这样盐便溶解在水里。然后在分离器中将油水分离,加入破乳剂有助于乳状液破乳,或使用更常用的方法,在沉降罐中用高电位电场来聚结两极凝集的盐水滴。脱盐过程中产生含油脱盐污泥和高温脱盐污水及其他过程中产生的废水都通过炼厂的污水处理设备一起排放。原油脱盐用水常常是来自炼制过程中其他未经处理或部分处理过的水。

常压蒸馏

　　脱盐原油通过换热器和高温加热炉加热至 750 ℉,进入常压立式原油分馏塔,大部分原料都会汽化并根据不同温度分别在 30~50 个塔盘上冷凝,将原油分成不同馏分。轻质馏分在蒸馏塔顶部冷凝聚集,重质馏分在常压塔中可能不会汽化,则需要进一步减压蒸馏。在每个常压蒸馏塔中,许多低沸点侧线馏分混合物(至少 4 个)将从不同塔盘上取出。同较重馏分一样,这些低沸点混合物也需要取出。这些侧线馏分被分别送入含有 4 ~ 10 个塔盘的不同小型汽提塔,每个汽提塔从最底层塔盘下面注入气流,气流从重组分中带出轻组分,两者回到相应侧线馏分抽出塔盘上方的汽提塔。常压蒸馏得到的馏分包括石脑油、汽油、煤油、轻燃料油、柴油、瓦斯油、润滑油馏分和釜底重质油,其中大部分都可以作为成品出售,或者和下游产品混合出售。常压蒸馏和其他炼油过程中的另一种产品是轻组分的、非冷凝炼制燃料气(主要是甲烷和乙烷),通常情况下,这种气体中还含有硫化氢和氨气。这些气体的混合物被称为酸性气体。酸性气体经炼厂酸性气体分离处理系统分离,可作为炼厂加热炉的燃料。常压蒸馏中排放的气体源于原油加热炉燃料的燃烧和其他一些短时排放。分馏塔也产生油性含硫污水(含硫化氢和氨的冷凝蒸汽)和油。

减压蒸馏

在常压蒸馏的温度和压力条件下不能分离的较重馏分要进行减压蒸馏。减压蒸馏是在较低压力(0.2 ~ 0.7psi,绝对压力)下进行石油分馏,促进轻组分挥发和分离。在大多数系统中,用蒸汽喷射泵、真空泵和气压冷凝器保持减压蒸馏塔内呈真空状态。在减压塔底部注入过热蒸汽是为了降低塔内碳氢化合物压力,加快汽化和分离。减压塔中的较重馏分可以继续通过裂化和焦化的下游过程加工成更有价值的产品。

原油蒸馏过程中,加热炉内燃料的燃烧及较轻气体从减压蒸馏塔冷凝器顶部散出时,可能会产生气体排放。一定量的不能冷凝的轻烃和硫化氢通过冷凝器到达热水井,然后排放到炼油厂含硫燃料系统或排放到加热器、火炬、其他控制设备以消除硫化氢。排放量取决于设备的大小、原料的类型和冷凝水的温度。如果减压塔使用大气冷凝器,会产生大量的含油废水。为消除此类含油污水,许多炼油厂已基本上用真空泵和表面冷凝器来取代大气冷凝器。减压分馏塔中也会产生油性含硫污水。

Key to Exercises

Ⅰ.1. The second phase is made up of three different types of "**downstream**" processes: combining, breaking, and reshaping. Supporting operations may include wastewater treatment, sulfur recovery, additive production, heat exchanger cleaning, blowdown systems, blending of products, and storage of products.

2. Before separation into fractions, crude oil usually must first be treated to

remove corrosive salts. The desalting process also removes some of the metals and suspended solids which cause catalyst deactivation.

3. The desalted crude oil is then heated in a heat exchanger and furnace to about 750 degrees (℉) and fed to a vertical, distillation column at atmospheric pressure where most of the feed is vaporized and separated into its various fractions by condensing on 30 to 50 fractionation trays, each corresponding to a different condensation temperature.

4. Fractions obtained from atmospheric distillation include naphtha, gasoline, kerosene, light fuel oil, diesel oils, gas oil, lube distillate, and heavy bottoms.

5. Heavier fractions from the atmospheric distillation unit that cannot be distilled without cracking under its pressure and temperature conditions are vacuum distilled.

Ⅱ. 1 – 5 DDDAA 6 – 10 BCBBD

Ⅲ. 1 – 5 ACBAB

Ⅳ. 1. 地质成因不同,原油的成分会有很大差异。

2. 炼油厂中最重要的一个工艺流程是先将原油蒸馏得到各种馏分。

3. 这些侧线馏分被分别送入含有 4～10 个塔盘的小型汽提塔,每个汽提塔从最底层塔盘下面注入气流。

4. 在大多数系统中,用蒸汽喷射泵、真空泵和气压冷凝器保持减压蒸馏塔内呈真空状态。

5. 如果减压塔使用大气冷凝器,则会产生大量的含油废水。

Ⅴ. 1. 石油炼制是一个复杂的过程,包括很多工艺过程,具体的工艺过程则取决于要炼制的原油和目标产品的性质。正因为这样,没有两家完全一样的炼油厂。部分产品从某些工艺过程中输出又返回,然后进入下一个工艺流程,又返回到上一个工艺,或与其他产品混合从而得到成品。

2. 原油分馏成各种馏分之前,通常先要脱掉腐蚀性的盐。脱盐过程也能除去一些导致催化剂失活的金属和固体悬浮物。脱盐过程包括将原油加热与水混合,这样盐便溶解在水里。然后在分离器中将油水分离,加入破乳剂有助于乳状液破乳,或使用更常用的方法,在沉降罐中用高电位电场来聚结两极凝集的盐水滴。

2.4 下游处理

导语

下游流程改变烃类化合物分子结构,把大个分子裂解成小分子,把小分子组合成大分子,或者重组分子以优化质量。处理方法包括热裂解、焦化、催化裂化以及加氢催化裂化等。现在催化裂化法已经在很大程度上取代了热裂解法,而流化焦化在不久的将来也将取代延迟焦化成为主要的工艺流程。

课文

原油蒸馏得到的某些馏分要通过各种方法进一步加工,这些方法包括热裂解(减黏裂化)、焦化,催化裂化、加氢催化裂化以及加氢精制、烷基化、异构化、聚合、催化重整、溶剂萃取、催化脱硫醇、脱蜡与丙烷脱沥青等。这些下游流程改变烃类化合物分子结构,把大个分子裂解成小分子,把小分子组合成大分子,或者重组分子、优化质量。下面提到的许多加工流程都应用了行业中一系列不同工艺。

1. 热裂解/减黏裂化

热裂解/减黏裂化过程是在较高温度与压力下将较大的烃分子分解为更小、更轻的分子。催化裂化法已经在很大程度上取代了热裂解法,一些炼油厂已不再使用热裂解法。这两种方法都能减少重质燃料油等低价值产品的生产,同时增加催化裂解的原料和汽油产量。在热裂解过程中,给料通常是重粗柴油和减压蒸馏后的渣油。

2. 焦化

焦化就是用裂解方法将炼厂产出的劣质重质燃料油转化为汽油、柴油等运输燃料的过程。给产品提纯升级的焦化过程还生产石油焦,主要由炭黑与不同数量的杂质组成。如果石油焦硫含量不高,可以作为发电厂的燃料使用。焦炭亦可作为非燃料使用,如作为许多炭和石墨产品的原料,包括生产铝电池的阳极,以及生产磷单质、钛白粉、碳化钙、碳化硅用的加热炉电极等。目前有许多不同工艺可用于生产焦炭,最广泛使用的是延迟焦化,但流化焦化预计在未来将成为一个主要工艺过程。流化焦化生产出的高质量的焦炭需求越来越大。延迟焦化过程与热裂化基本一样,只不过进料反应时间较长,不用冷却。上游各工序的釜底重质油馏分是延迟焦化的进料,这些馏分首先进入分馏塔,较轻组分取出,重组分凝聚于底部。之后重组分移入加热炉中并加热至 900 ~ 1000 °F,接下来转入称为焦化塔的绝热容器中,形成焦炭。当焦化塔填满时,进料就转移到一个空的并行塔中。从焦化塔出来的热蒸气含有较轻的烃类产品,例如硫化氢和氨,这些热蒸气要返回到分馏塔由酸气处理系统处理,或作为中间产品取出。之后将水蒸气注入焦炭塔中以移出较少烃蒸气,注入水冷却焦炭并将焦炭移出。通常情况下,用高压水射流将焦炭从焦化塔中分离出来。

焦化流程的空气污染物排放包括加工热水器烟气排放、短时排放及在焦化塔分离焦炭过程中的排放。注入的蒸汽冷凝,剩余的气体一般都排放烧掉了。而废水是在分离焦炭、冷处理以及蒸汽注入过程中产生的。此外,从焦化塔取出焦炭可导致微粒排放以及剩余的烃组分排放到大气中。

3. 催化裂化

催化裂化是在高温高压和催化剂共同作用下将较大的烃分子分解成更小、更轻的分子的过程。由于可以生产更多的高辛烷值汽油并减少重质燃料油和轻烃的产出,催化裂化已经在很大程度上取代了热裂解。催化裂化原料主要来源是原油蒸馏过程中得到的轻质和重质馏分油,目的主要是加工成汽油和一些燃料油及轻烃。催化裂化中大部分情况下使用的催化剂包括合成结晶硅铝,称为沸石,以及非晶合成硅—铝的混合物。催化裂化过程以及炼油厂其他大多数催化过程中生产的焦炭会聚集在催化剂表面,降低其催化活性,因此需要连续或定期将附着在催化剂表面的焦炭高温烧掉而保持催化剂活性。催化剂再生的方法和频率都是催化裂化装置设计时需考虑的主要因素。目前美国正在使用的一些催化裂化工艺设计包括固定床反应器、移动床反应器、流化床反应器以及直流炉燃烧器。

在流化床过程中,已预热至 500~800 ℉的石油和石油气在反应器中或至反应器的进料管中与热催化剂在 1300 ℉下接触。该催化剂呈细粒状,在和水蒸气混合时具有流体的很多特性。流化催化剂和反应器中已反应的轻烃蒸气要进行机械分离,催化剂上的残余油品经水蒸气汽提带出。裂解后的油蒸气再输送回分馏塔,分离成各种所需的组分。催化剂流入另一个容器中,在空气中燃烧掉积炭,实现一次或两次催化剂再生。

在移动床过程中,油加热到 1300 ℉,反应器加压,然后油接触催化剂珠或颗粒流,之后裂解的产物进入到分馏塔分离得到产品。只要催化剂积炭燃烧掉,催化剂便可连续再生。有的装置还使用蒸汽汽提残余氧气和烃类,然后再让催化剂回到油蒸气中。

4. 催化加氢裂化

催化加氢裂化通常使用固定床催化裂化反应器,在有氢气且压力高(1200~2000psi,表压)的条件下进行裂化。加氢裂化装置通常用于处理那些难以裂解以及不能在催化裂化装置中有效裂化的产物,包括中间馏分油、循环油、残余燃料油和拔顶油。氢抑制残余重质成分形成,通过与裂化产品反应增加汽油产量。然而这个过程也会裂解含硫和氮的重质烃类化合物,释放杂质,这些杂质会对催化剂活性造成潜在的影响。因此,原料往往首先加氢处理,除去杂质,然后再进行催化加氢裂化。有时用加氢反应过程的第一个反应器完成加氢处理,除去杂质。水对加氢裂化催化剂也有一些不利影响,原料进入反应器之前必须脱水。原料经过进料管时,硅胶或分子筛网烘干机将水脱除。根据所需产品和装置的大小,催化加氢裂化在单级或多级反应器中进行。大多数催化剂是由硅—铝与少量稀土金属结晶混合制成的。

加氢裂化原料通常需要先进行加氢处理,去除对催化剂不利的硫化氢和氨。分馏塔会产生酸性气体和含硫污水,如果加氢裂化原料首先经过加氢处理除去杂质,那么酸气和含硫污水中硫化氢和氨的含量都会降低。加氢裂化催化剂通常是在使用2~4年后在厂区外进行再生的,因此,再生过程中很少或不会产生排放。空气排放物主要源于加热器、通风孔和短时排放。

Key to Exercises

Ⅰ. 1. Certain fractions from the distillation of crude oil are further refined in thermal cracking (visbreaking), coking, catalytic cracking, catalytic hydrocracking, hydrotreating, alkylation, isomerization, polymerization, catalytic reforming, solvent extraction, merox, dewaxing, propane deasphalting and other operations.

2. A number of different processes are used to produce coke; "delayed coking" is the most widely used today, but "fluid coking" is expected to be an important process in the future. Fluid coking produces a higher grade of coke which is increasingly in demand. In delayed coking operations, the same basic process as thermal cracking is used except feed streams are allowed to react longer without being cooled.

3. The delayed coking feed stream of residual oils from various upstream processes is first introduced to a fractionating tower where residual lighter materials are drawn off and the heavy ends are condensed. The heavy ends are removed and heated in a furnace to about 900 ~ 1,000 degrees (℉) and then fed to an insulated vessel called a coke drum where the coke is formed.

4. Air emissions from coking operations include the process heater flue gas emissions, fugitive emissions and emissions that may arise from the removal of the coke from the coke drum. The injected steam is condensed and the remaining vapors are typically flared. Wastewater is generated from the coke removal and cooling operations and from the steam injection. In addition, the removal of

coke from the drum can release particulate emissions and any remaining hydro-carbons to the atmosphere.

5. In the fluidized – bed process, oil and oil vapor preheated to 500 to 800 degrees (℉) is contacted with hot catalyst at about 1,300 (℉) either in the reactor itself or in the feed line (riser) to the reactor. The catalyst is in a fine, granular form which, when mixed with the vapor, has many of the properties of a fluid. The fluidized catalyst and the reacted hydrocarbon vapor separate mechanically in the reactor and any oil remaining on the catalyst is removed by steam stripping. The cracked oil vapors are then fed to a fractionation tower where the various desired fractions are separated and collected. The catalyst flows into a separate vessel(s) for either single-or two-stage regeneration by burning off the coke deposits with air.

In the moving – bed process, oil is heated to up to 1,300 degrees (℉) and is passed under pressure through the reactor where it comes into contact with a catalyst flow in the form of beads or pellets. The cracked products then flow to a fractionating tower where the various compounds are separated and collected. The catalyst is regenerated in a continuous process where deposits of coke on the catalyst are burned off. Some units also use steam to strip remaining hydrocarbons and oxygen from the catalyst before being fed back to the oil stream.

Ⅱ.1－5 BACCB 6－10 ACBCA

Ⅲ.1－5 ACACA

Ⅳ.1. 在热裂解过程中,给料通常是重粗柴油和减压蒸馏后的渣油。

2. 当焦化塔填满时,进料就转移到一个空的并行塔中。

3. 注入的蒸汽冷凝,剩余的气体一般都排放烧掉。

4. 催化裂化是在高温高压和催化剂共同作用下将较大的烃分子分解成更小、更轻分子的过程。

5. 因此,需要连续或定期将附着在催化剂表面的焦炭高温烧掉而保持催化剂活性。

Ⅴ.1. 热裂解/减粘裂化过程是在较高温度与压力下将较大的烃分子分解为更小、更轻的分子。催化裂化法已经在很大程度上取代了热裂解法,一些炼油厂已不再使用热裂解法。这两种方法都能减少重质燃料油等低价值产品的生产,同时增加催化裂解的原料和汽油产量。在热裂解过程中,给料通常是重粗柴油和减压蒸馏后的渣油。

2. 加氢裂化原料通常需要先进行加氢处理,去除对催化剂不利的硫化氢和氨。分馏塔会产生酸性气体和含硫污水,但如果加氢裂化原料首先经过加氢处理除去杂质,那么酸气与含硫污水中硫化氢和氨的含量都会降低。加氢裂化催化剂通常是在使用两年到四年后在厂区外进行再生的。因此,再生过程中很少或不会产生排放。空气排放物主要源于加热器、通风孔和短时排放。

第3章 一次加工

3.1 蒸馏

导语

石油炼制始于原油的物理分离,即将原油蒸馏分离成不同烃类。原油在常压和减压蒸馏塔中分离为不同沸程的馏分。蒸馏塔和裂解塔不同,了解蒸馏塔结构和蒸馏机理对于理解原油的一次加工至关重要。

课文

在炼厂中,人们很容易将很多高的塔误认为是裂解塔。实际上,大部分的高塔都是不同种类的蒸馏塔,裂化塔大都比较矮粗。

蒸馏装置是工艺工程师根据蒸馏曲线的重要特性而做出的巧妙发明。其实,它们使用的原理并不复杂。但为了便于读者完整地了解和熟悉其原理,本文在这里将进行基本讲解。

简单蒸馏

很多年来,肯塔基州的酿酒师们就利用图3.1所示的简单蒸馏装置将威士忌酒从酒糟中分离出来。例如,将酸性麦芽浆发酵后,部分麦芽浆会慢慢地经历一个化学变化过程形成酒精,将酒精加热到沸点温度后,自制的威士忌酒就会汽化形成蒸气。由于气体比液体的密度小,蒸气就可以从液体中溢出,经过冷却器冷凝重新变为液体。蒸馏器中剩余的液体就废弃掉,冷却器中的液体则为成品。工艺工程师们称这个过程为简单的间歇式蒸馏。

如果想得到比一般产品质量更好的酒,酿酒师们可能会按照第一次蒸馏的方法将产品进行二次蒸馏。二次蒸馏中可能会从非酒精杂质中分离出最好的成分。非酒精杂质是在第一次蒸馏中进入塔顶。杂质流向塔顶可能是由于温度控制不够精确,也可能是因为造酒师为了增加产量故意提高蒸馏塔温度而造成的。

这种两步式操作可以按照图3.2所示的方法变为连续蒸馏。事实上,很多早期的石油蒸馏操作就是这样实现的。

蒸馏塔

很明显,这种两步间歇蒸馏操作无法处理日均几十万桶的原油,而且原

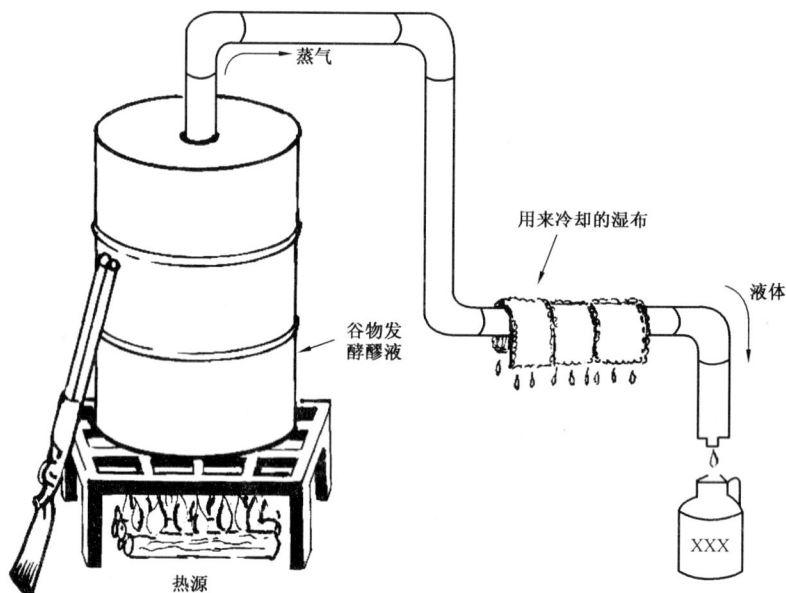

图3.1　酿酒用的蒸馏釜

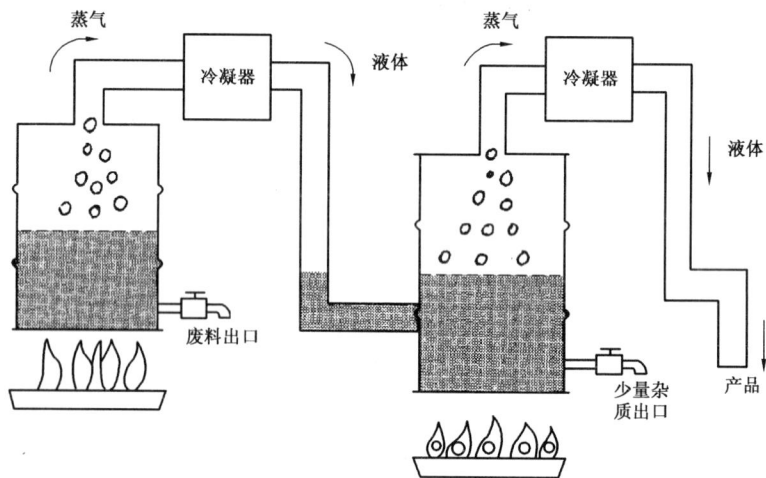

图3.2　两级间歇蒸馏

油中有五种到六种组分需要分离。但是蒸馏塔可以通过连续进料实现这一目标,而且蒸馏塔需要的设备更少,消耗的人力和能量也更少。

　　图3.3 显示了原料进入原油蒸馏塔前、后的变化。原油进入蒸馏塔,得到气体(丁烷和轻质烃)、汽油、石脑油、煤油、轻瓦斯油、重瓦斯油和渣油等产品。

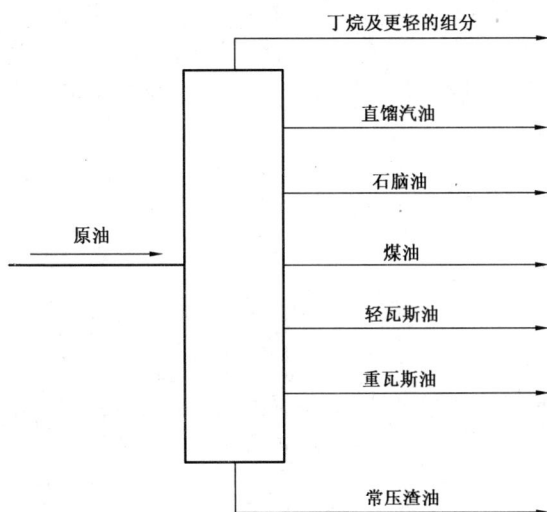

图 3.3　原油蒸馏产物

蒸馏塔内部的加工过程更为复杂。加料泵是蒸馏操作最重要的设备,将原油从储油罐中泵出(图 3.4)。原油泵入熔炉,加热到约 750 ℉ 的温度。当炉温升至这一温度时,一半以上的原油可以转化成气体。然后这种气液混合物会进入蒸馏塔中。

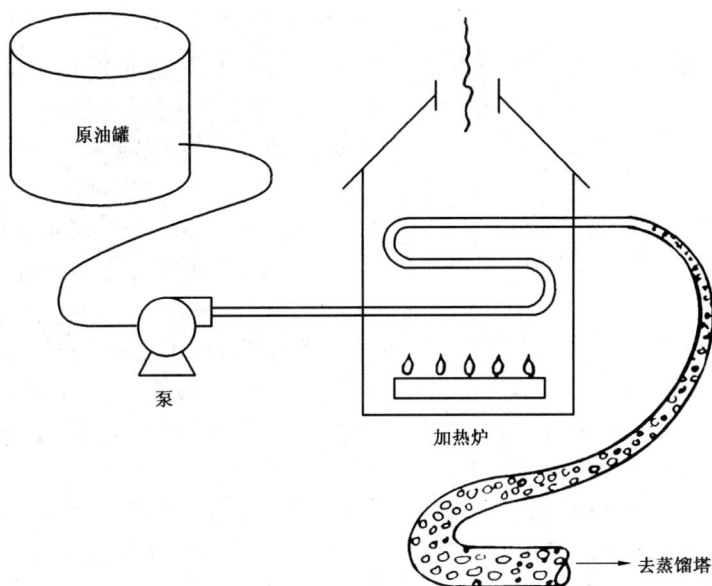

图 3.4　蒸馏前原油的预处理

在蒸馏塔中有一系列塔板,板上有孔。蒸气可通过这些孔上升流过蒸馏塔。当原油气液混合物进入蒸馏塔中时,重力作用就使得重质液体组分流向塔底,轻质气体组分通过塔板到达塔顶,如图3.5所示。

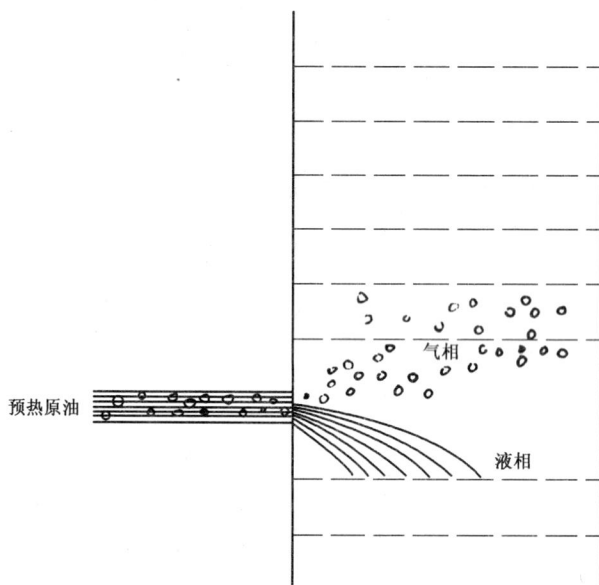

图3.5　原油进入蒸馏塔示意图

塔板的孔眼上有一种称为泡罩的构件(图3.6)。它们的作用就是使通过塔板的蒸气经过板上几英尺深的液体后变成气泡。鼓泡是蒸馏操作的重要环节,即热蒸气通过液体形成气泡(750 °F以上)。鼓泡时蒸气的热量可以传给液体。当气泡温度降低时,气泡中的一些碳氢化合物就从气态变为液态。蒸气温度降低,液体温度也降低,使蒸气中的重组分浓缩形成液体。

通过液体后,蒸气中的重组分分离出去,气体继续流向上层塔板,重复相同的过程。

与此同时,由于气体中的一些重组分都形成了液体,因此每块塔板上液体的量逐渐增加。图3.7显示了一种称为降液管的装置,其作用就是使过多的液体通过降液管溢流到下层塔板。在蒸馏塔的某些位置,液体馏分通过侧线采出,轻组分从塔顶采出,重组分从塔底采出。

实际上,一些组分会经历数次循环蒸馏。气体穿过塔板上升,最后冷却形成液体,液体通过降液管流向下层塔板。这种气液互相洗涤的方法可以将那些不能一次分离的馏分分离出去。

回流和再沸

在蒸馏塔外也要做一些工作,以使蒸馏操作顺利进行。为了保证一些

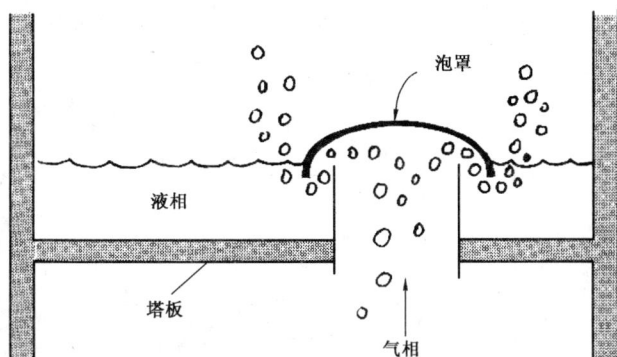

图 3.6　蒸馏塔中的泡罩阀

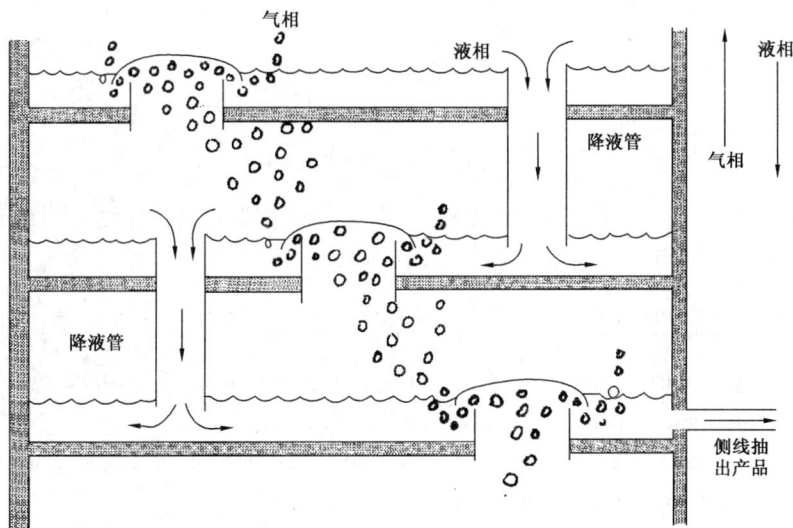

图 3.7　降液管及侧线抽出

重组分不从塔顶溢出,有时候气体需要冷凝。冷凝的液体将会回流到下面的塔盘中,剩余的气体作为产品放出。这个过程就是回流(图 3.8)。

　　反过来,在塔底的液态产品也会包含部分轻质烃。可用侧线抽出并重新在加热炉循环这些液体,使轻烃发挥出来再回到较低层的塔板上,这个过程称为再沸。

切割温度

　　对蒸馏过程进行分析和控制,关键参数是切割温度,即各种蒸馏产品分离的温度。产品(馏分)开始沸腾的温度称为初馏点,产品全部蒸发完的温度称为终馏点。因此,每次切割都有两个切割温度,即初馏点和终馏点。石

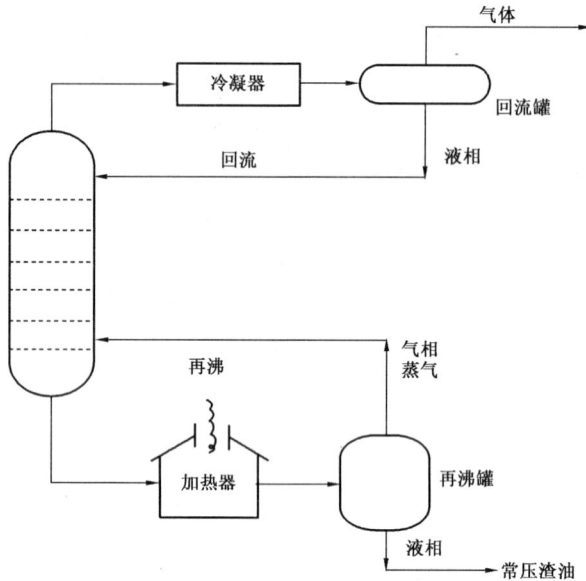

图3.8 再沸和回流

脑油的终馏点正好是煤油的初馏点。在某一切割温度点,两种相邻组分的初馏点和终馏点重合。

Key to Exercises

Ⅰ.1. A casual passerby of a refinery can make an easy mistake by referring to the many tall columns inside as "cracking towers".

2. For years Kentucky moonshiners used the simple still in Fig. 3. 1 to separate the white lightning from the dregs.

3. After the sour mash fermented—i. e. , a portion of it had slowly undergone a chemical change to alcohol—they heated it to the boiling range of the alcohol. The white lightning vaporized as a vapor, it is less dense (lighter) than liquid. It moved out of the liquid, then through the condenser where it cooled and turned back to liquid. The liquid left in the still was discarded. The liquid that ended up in the condenser was bottled.

4. To assure some of the heavies don't get out the top of the column, occasionally some of the vapor will be run through a cooler.

5. For analyzing and controlling distilling operations, the key parameters are cut points, the temperatures at which the various distilling products are separated.

Ⅱ.1－5　ABABC　6－8　DBC

Ⅲ.1－5　ABBBD

Ⅳ.1. 蒸馏装置是工艺工程师根据蒸馏曲线的重要特性而做出的重大发明。

2. 很明显,这种两步间歇蒸馏操作无法处理日均几十万桶的原油,而且原油中有五种到六种组分需要分离。

3. 原油进入蒸馏塔后可以直接得到气体(丁烷和轻质烃)、汽油、石脑油、煤油、轻瓦斯油、重瓦斯油和渣油等产品。

4. 与此同时,由于气体中的一些重组分都形成了液体,因此每块塔板上液体的量逐渐增加。

5. 相应的,在塔底的液相产品也包含有部分轻烃产品。

Ⅴ.1. 实际上,一些组分会经历数次循环蒸馏。气体穿过塔板上升,最后冷却形成液体,液体通过降液管流向下层塔板。这种气液互相洗涤的方法可以将那些不能一次分离的馏分分离出去。

2. 反过来,在塔底的液态产品也会包含部分轻质烃。可用侧线抽出并重新在加热炉循环这些液体,使轻烃挥发出来再回到较低层的塔板上,这个过程称为再沸。

3.2　减压闪蒸

导语

在蒸馏过程中,所有难以控制的裂化现象都可能出现。对温度敏感的物质,只有减压蒸馏才能不被破坏。减压蒸馏就是在低于常压的条件下低温蒸馏。为避免这些现象发生,炼厂研发出减压闪蒸技术。

课文

迄今为止,关于蒸馏曲线和蒸馏塔,一般关注的是温度为 900 ℉ 左右的情况,在该温度附近所发生的现象称为裂化。炼厂虽然广泛使用裂化技术,但还是在一定限度内。

低压的影响

假如你拥有一个专卖烤马铃薯的连锁饭店——一个在得克萨斯州的休斯敦,一个在科罗拉多州的丹佛。你知道在丹佛烤马铃薯的时间比在休斯敦要多花 20 分钟吗?(当我在落基山脉滑雪时,一天晚上自己做晚饭发现了这个现象)

原因在于,丹佛海拔较高,因而气压较低。当人们说高山上空气更稀薄时,他们其实是想说每立方英寸氧气比在山下的要少。由于压力降低,水的沸点也随之下降。即使将炉温设定在 375 ℉,但在丹佛水蒸气将使马铃薯温度保持在 205 ℉,而不是休斯敦的 212 ℉。因此,在丹佛烤出的马铃薯不如休斯敦热,从而需要烤更长时间。

上面阐述了温度和压力之间的简单关系。加热过程使得分子吸收足够的能量,从液态变为气态。逸出速率取决于热量传输速率(炉温)和液体上部空气的压力。压力越低,所需传递的能量越少,因此从液相中逸出蒸气(如鼓泡)的起始温度越低。因此,压力越低,沸点越低。

减压闪蒸过程

现将压力和沸点之间的简单关系应用到原油裂解问题上来。如果温度过高,常压直馏渣油将会发生裂解反应,但常压直馏渣油需要分馏出多种馏分。因此可使用减压蒸馏来解决这一问题。

在图 3.9 所示的真空闪蒸罐中,对来自蒸馏装置的直馏渣油,在温度依然较高的状态下,用泵打入到闪蒸罐中。闪蒸罐是大直径容器,由于其外形低矮,难以称为塔,但多少也算个塔。直馏渣油离开泵后一路向下进入闪蒸罐,此时其压力已降低到常压以下。在那样的低压下,渣油中的部分轻组分汽化,但在该温度下不会发生裂化反应。

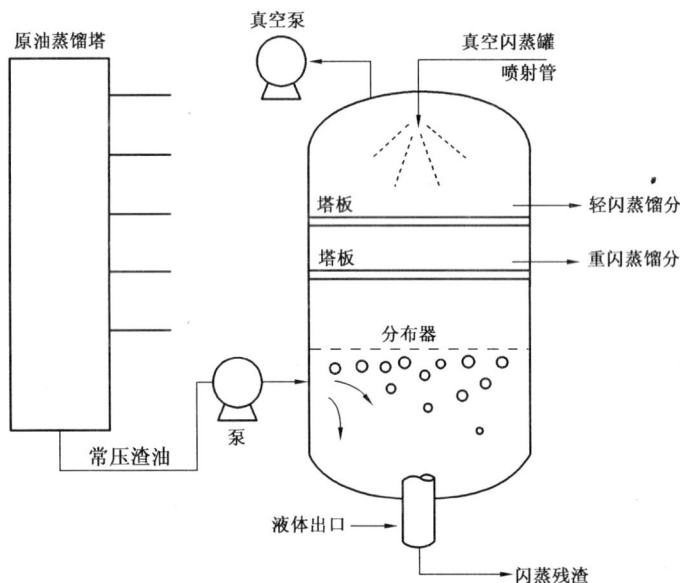

图 3.9　减压闪蒸

闪蒸罐内的各种构件用于分离并收集排出物。气液混合物进入到闪蒸罐中时,几乎所有的液体落到罐底。其中的一部分液体形成液滴——不是完全气态,但夹带在直馏渣油的汽化物中,开始随蒸气上升。为了获取这些液滴并使蒸气均匀分散于闪蒸罐内,气液混合物需经过一个分布器——一个厚的网筛或者是一个几英寸厚带有稀疏金属环的塔板。分布器捕捉液滴,使其落到底部。闪蒸罐中再往上是两层塔板,其工作机制与蒸馏塔的一样。闪蒸罐也可以有一个再沸器,不过图中没有显示出来。回流与蒸馏塔的作用一样,将一部分从塔板中抽出的冷却液流从闪蒸罐的顶部向下喷洒。闪蒸罐顶部的真空泵使罐内处于低压状态,并不断地抽出没有冷凝的气体,这些气体通常含有少量水和一些烃。有些闪蒸罐用蒸汽喷射器来保持低压环境。

从闪蒸罐的上部抽出两股物流:轻闪蒸馏分和重闪蒸馏分,两股馏分通常是分开的,统称为闪顶产品。炼油工作者通过化学分析发现,闪顶产品中重闪蒸馏分里含有污染物,会极大地毒害下游加工装置中的催化剂。两股物流处于分离状态,便可对重闪蒸馏分进行处理,除去大部分有害物质。

同时,在闪蒸罐底部,液体通过一根管子排出。有些闪蒸罐有用于最后一次分离底部物流中有用组分的设备。通常是在液体流出时,通过管子向上泵入超热蒸汽,释放出液体中可能含有的轻烃。因为超热蒸汽温度远高于裂化温度,从底部流出的液体(闪蒸残渣,也称减压渣油)在还未发生裂化反应前就与冷液流相遇,急剧冷却下来(裂化反应至少需要一些停留时间才能开始反应)。

闪蒸残渣有几个用途——作为沥青、热裂化、焦化装置的进料或残渣燃料油的调和组分。闪蒸的真正原因是用闪顶产品来生产催化裂化的原料。

评述

闪蒸等同于在 1000~1100 ℉的温度范围内对直馏渣油进行蒸馏。原油的蒸馏曲线总是显示温度和体积的关系,理论上蒸馏似乎是这样的,也就是说,研究者认为真空闪蒸塔是蒸馏装置的一部分。

Key to Exercises

Ⅰ.1. There is a phenomenon that happens at these temperatures called cracking.

2. Since the pressure is lower, water will boil at a lower temperature.

3. The straight run residue will crack if the temperature goes too high, but straight run residue needs to be separated into more cuts.

4. The two streams drawn off the upper part of the flasher, light flashed distillate, and heavy flashed distillate, together are called flasher tops.

5. The real reason to run a flasher is to make the flasher tops to feed to a

cat cracker.

Ⅱ. 1 - 5 CABAB 6 - 8 ADD

Ⅲ. 1 - 5 BAADC

Ⅳ. 1. 原因在于丹佛海拔较高,因而气压较低。

2. 上面阐述了温度和压力之间的简单关系。加热过程使得分子吸收足够的能量,从液态变为气态。

3. 随着气液混合物蒸气进入到闪蒸罐中,几乎所有的液体集中到罐底。

4. 炼厂广泛使用裂化技术,但受到控制条件的制约。

5. 同时,在闪蒸罐底部液体通过一根管子排出。有些闪蒸罐有用于最后一次分离底部物流中有用组分的设备。

Ⅴ. 1. 现将压力和沸点之间的简单关系应用到原油裂解问题上来。如果温度过高,常压直馏渣油将会发生裂解反应,但是常压直馏渣油需要裂化分解出多种的馏分。因此使用减压蒸馏来解决这一问题。

2. 闪蒸残渣有几个用途——作为沥青、热裂化、焦化装置的进料或残渣燃料油的调和组分。闪蒸的真正原因是用闪顶产品来生产催化裂化的原料。

3.3 芳烃采收与脱硫

⊞ 导语

芳香族烃可以是单环芳烃或多环芳烃,单环芳烃只含一个由 6 个碳原子构成的苯环。"芳烃"这一术语在其物理机制被发现之前就有了,是因为许多化合物有香甜的气味。原油中的硫表现为最简单化合物硫化氢,或者复杂的环结构。芳烃和硫都会破坏混合物的质量,而且芳烃分离后其价值更高,因此有必要采收芳烃并脱硫。

⊞ 课文

芳烃采收

去除芳烃化合物的原因有两个,其中一个原因是芳烃的存在会对混合物的质量产生不利影响,另一个原因是芳烃分离后其价值更高。下面是一些例子:

(1)汽油有最大苯含量的限制。

(2)煤油中芳烃化合物的存在会导致烟点不合格。

其他方面的应用包括:

（1）不含芳烃和含有芳烃的煤油溶剂工业应用不同。

（2）将芳烃中的苯、甲苯、二甲苯分离出来，可用于化工生产的许多方面。

（3）从重瓦斯油馏分中脱除芳烃，可以改善润滑油的性能。

工艺

溶剂采收基本原理是某溶剂能选择性地溶解其他种类的化合物。据此，某些溶剂可以只溶解芳烃而不溶解烷烃、烯烃及环烷烃。对该工艺的具体原理在此不再叙述。

溶剂采收工艺首先要求溶剂与萃取物易于从油品中分离；其次要求溶剂和萃取物在分馏塔中易于分离。

以含有较多芳烃组分的煤油为例，在烧杯中加入一半煤油和一半溶剂，如液态二氧化硫。混合后液体分为两相，下部是煤油，上部是液态二氧化硫，但煤油体积将少于一半，而液态二氧化硫由于溶解了芳烃，体积大于半个烧杯。

将液态二氧化硫倒出，再通过简单蒸馏即可把溶解的芳烃分离出。该两步法称为间歇式溶剂采收工艺。

了解了简单间歇式工艺的原理，对于连续流水作业也就易于形成概念。如图3.10所示的三塔系统，原料呈气态从萃取塔底部进入，塔内有布置得如迷宫般的混合器（有时混合器可以做机械运动，如在旋转接触器内），溶剂以液态形式从萃取塔顶部加入，溶剂在向下流动的过程中发挥作用，溶解与它接触的萃取物。剩余的烃类上升至萃取塔顶部，称为萃余相。

两个蒸馏塔用于处理从萃取塔中出来的物流，其中一个清除萃余相中携带的剩余少量溶剂，另一个则将溶剂与萃取物分离。由这两个塔采收的溶剂进入萃取塔顶部循环使用。

采收不同芳烃所用溶剂如下：

（1）煤油处理：液态SO_2，糠醛。

（2）润滑油处理：液态SO_2，糠醛，苯酚，丙烷（分离烷烃与沥青质）。

（3）汽油处理：环丁砜，苯酚，乙腈，液态SO_2。

苯和芳烃的采收

应用溶剂萃取最多的地方就是对混合芳烃的采收，特别是对苯的采收。为使工艺更有效，需提取重整油或直馏汽油中的中段馏分，得到芳烃浓缩物作原料，如图3.11所示。由于芳烃浓缩物含有较高的苯含量，萃取过程会更有效。

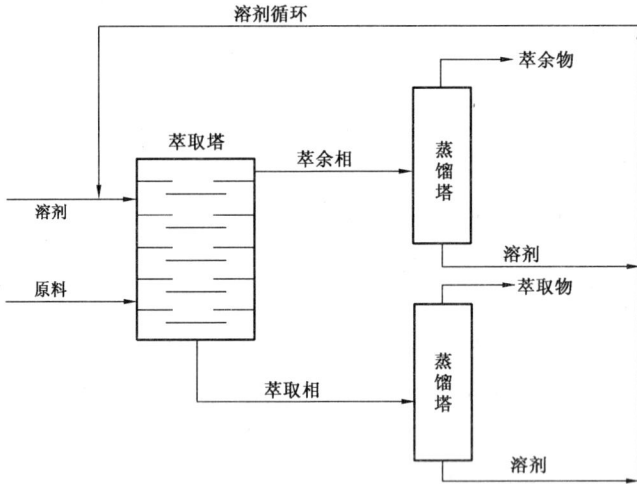

图 3.10　溶剂回收过程

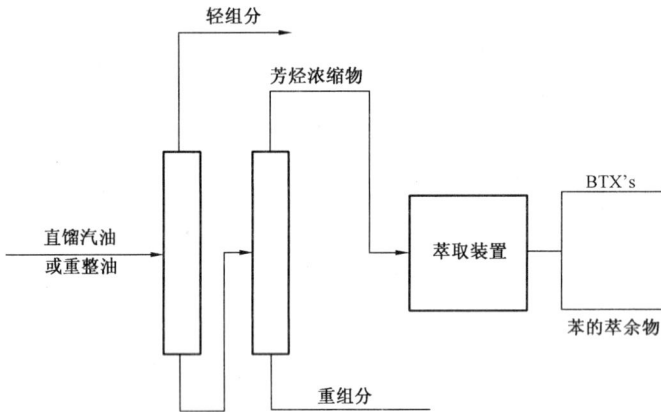

图 3.11　BTX 回收

一些名词的命名经常会令人误解,如苯萃余物就不含有苯,它是芳烃浓缩物脱除苯后的剩余物。最终,苯的萃余物成为汽油调和组分,而分离得到的苯则成为化工原料。

脱硫

原油中含有硫,其存在形式多种多样,既有最为简单的化合物硫化氢,也有复杂的环状结构。原油蒸馏过程中,分解较高沸点的硫化物会产生硫化氢,硫化氢会进入液化石油气里,由于硫化氢有毒性和腐蚀性,所以必须将其从液化石油气中除去。方法是,用胺溶液进行逆流洗涤,在另外的容器里加热该胺溶液脱硫,去除硫化氢,再次得到胺溶液循环用于逆流洗涤。硫醇可以看做是硫化氢的衍生物,即硫化氢中氢原子被碳氢结构替代了。那些沸点在80℃以下的硫醇极容易溶解于碱性溶液,但当温度超过80℃时,其溶解度就会迅速减小。因此,在生产液化石油气和轻质汽油时,会采用苛性钠溶液进行逆流洗涤,以除去硫醇。

沸点在80～250℃之间,馏分里的硫醇无法萃取,但是能在混有空气的梅洛克斯溶液里氧化为二硫化物,这些二硫化物无腐蚀性,几乎无味,且依旧溶解在油中。另一种使硫醇氧化的方法是使用氯化铜作催化剂。上述两种工艺均可用于加工喷气式发动机燃料。

当蒸馏原油所得馏分的沸点越来越高时,馏分的含硫量也在增加。生产柴油和家庭供暖系统燃料油的沸点在250～350℃之间,在这个沸点范围内,大部分中东产的原油含硫量大约为1%(质量分数)。当这种燃料燃烧时,硫就氧化成二氧化硫,而二氧化硫又极易氧化成硫酸,会引起大气污染和金属腐蚀。由于硫主要与碳、氢结合,其结构比简单的硫醇复杂得多,它就不能用上面介绍的方法处理。这些复杂的化合物必须经过裂解才能得到硫。具体方法如下:在高温(320～420℃之间)条件下,使原油连同氢气一起通过以铝矾土为载体的氧化钴和氧化钼催化剂(催化剂要制成小球或压制物)。当氢气与进料的比例超过反应本身所需比例几倍时,该反应会更容易发生,催化剂的寿命也较长。在这些条件下,硫化物就会分解;硫与氢气化合生成硫化氢。几乎所有的硫化物都可以用这种方法进行分解而不影响其他的碳氢化合物。

Key to Exercises

Ⅰ.1. Either the aromatics have detrimental effects on the quality of the mixture they're in, or the aromatics are worth more if they're separated than if they're not.

2. It is based on the ability of certain compounds to dissolve certain classes of other compounds selectively.

3. The aromatic compounds can be "sprung" from it by simple distillation.

4. The feed to the process is pared down to an aromatics concentrate by making a heart cut from a reformate or straight run gasoline stream. The aromatics concentrate then has a large benzene content, making the extraction process more efficient.

5. Mercaptans in fractions boiling between 80 and 250℃ cannot be extrac-

ted but can be oxidized to disulphides in the Merox solution with air. The disulphides, which are no – corrosive and have little smell, remain dissolved in the oil. Another process for the oxidation of mercaptans uses copper chloride as a catalyst.

Ⅱ.1 – 5　BACDB　6 – 8　ACB

Ⅲ.1 – 5　AABDB

Ⅳ.1. 不含芳烃或含有芳烃的煤油溶剂工业应用各不相同。

2. 溶剂回收工艺首先要求溶剂与萃取物易于从油品中分离。其次要求溶剂和萃取物在分馏塔中易于分离。

3. 其中一个塔用于清除萃余相中携带的剩余少量溶剂。

4. 一些名词术语的命名经常会产生误导,如苯萃余物就不含有苯。

5. 原油中含有硫,其存在形式多种多样,既有最为简单的化合物 H_2S,也有复杂的环状结构。

Ⅴ.1. 溶剂采收基本原理是某溶剂能选择性地溶解其他种类的化合物。据此,某些溶剂可以只溶解芳烃而不能溶解烷烃、烯烃及环烷烃。该工艺的具体原理在此不再叙述。

2. 沸点在 80～250℃之间,馏分里的硫醇无法萃取,但是能在混有空气的梅洛克斯溶液里氧化为二硫化物,这些二硫化物无腐蚀性,几乎无味,且依旧溶解在油中。另一种使硫醇氧化的方法是使用氯化铜作催化剂。上述两种工艺均可用于加工喷气式发动机燃料。

第4章 二次加工

4.1 催化裂化(1)

🔲 导语

为了满足现代社会对汽油越来越多的需求,裂解技术的发明层出不穷,如轻裂解、热裂解和催化裂解等,其中最为广泛应用的是催化裂化方法。催化裂化的目的是将重质馏分油转化为汽油和更轻的组分。催化剂及其反应装置的逐步改进使裂解效果越来越好。

🔲 课文

在石油工业发展的早期,汽油需求的增长速度远快于对燃料油的需求增长。对炼油厂来说,加工原油生产直馏汽油以满足市场的同时,会造成燃料油的过剩。而这将彻底改变19世纪为生产足够的燃料油造成汽油过剩的局面。在新的经济形式下,汽油的价格逐渐上涨而重质组分的价格逐渐下降。

为了解决这一现实经济问题,极具创造力的工程师们开发了一系列裂化技术,其中最著名的是催化裂化技术。

工艺简介

催化裂化的过程为:在催化裂化反应器中,直馏重馏分油在一定的温度、压力条件下与催化剂接触以促进裂化反应。

催化剂的定义:催化剂是这样一种物质,它能够促进或引起化学反应,而化学反应前后其本身不发生化学变化,它能引起其他物质发生反应。催化裂化装置由三部分组成:反应器、再生器和分馏塔。

反应器

催化裂化装置的核心部分是反应器(图4.1),反应器由原料预热装置和提升管反应器(包括提升管和沉降器)组成。预热装置将原料加热至900～1100 ℉,然后原料和催化剂混合后一起进入提升管反应器。水蒸气随着催化剂一起进入,目的是为提升管内的混合物料提供足够的上升力,从而将物料抬升到分离器底部。上述反应都在提升管内进行。尽管物料在提升管内的停留时间只有短短数秒,但在高温条件下,物料也能与催化剂充分接触。

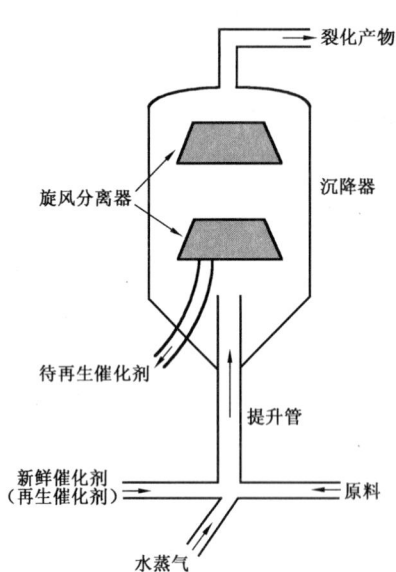

图 4.1　反应器

在老式的催化裂化装置中,沉降器称为反应器,因为大部分反应在其中进行。而在新式的催化裂化装置中,沉降器只是将催化剂从产物中分离出来。

原料和催化剂混合物从提升管出来以后进入旋风分离器,旋风分离器带动混合物高速旋转。因为催化剂的密度大,受离心力的作用,催化剂撞击到旋风分离器壁后沿器壁滑落到底部,通过重力的作用脱离沉降器。原料大部分以气体的形式,少部分以小液滴的形式从旋风分离器中出来后,上行进入另一个旋风分离器,进一步分离净化后从上出口离开沉降器。

为了延长原料与催化剂的接触时间,工程师们试图设计一种反应装置,使反应在提升管内而不是在沉降器内进行。但是在一些提升管催化裂化装置中,一部分原料可以从提升管中部进入,从而减少接触时间。这是因为成分不同的原料与催化剂的反应速率不同,这种进料方式可以处理不同的原料油。

催化剂

现今所使用的催化裂化催化剂可以说是一项伟大的革命。早期的催化剂由天然白土制成,现在使用的催化剂是分子筛,催化性能有很大提高。分子筛催化剂有3种特性。如果将一瓶分子筛催化剂摇晃一下,就会发现,奶粉一样的白色粉末像液体一样流动。这种特性在整个工艺的设计过程中非常重要,并因此衍生出另一个术语——流化催化裂化(FCC)。

催化裂化催化剂的第二个特性是裸眼观察下不透明。而在显微镜下,会发现催化剂颗粒上有大量的孔,因而具有很高的表面面积。如果催化剂颗粒像地球一样大,那么催化剂的孔道比科罗拉多大峡谷还要深并且每隔几千米就有这样一个大峡谷。催化剂对反应的影响主要取决于催化剂与原料的接触程度,因此表面面积大在反应过程中非常重要。

第三个特性来自于技术的飞跃。早期白土催化剂含有促进裂化反应必需的天然矿物质,现在可以合成具有精确尺寸和矿物含量的催化剂。催化剂的孔道可以精确设计制作,只允许特定尺寸的分子通过,一次一个,这样可以控制特定类型分子的反应方式。催化剂供应商可以提供各种催化剂以满足不同要求,如有利于生产高辛烷值汽油组分,提高轻油收率,可以使用重质原料,降低积炭,降低反应温度以节省能耗等。

Key to Exercises

Ⅰ.1. It became apparent to refiners that distilling enough crude to make straight run gasoline to satisfy the market would result in a glut of fuel oils.

2. In a cat cracker, straight – run heavy gas oils are subjected to heat and pressure and are contacted with a catalyst to promote cracking.

3. The cat cracker comprises the reaction section, the regenerator and the fractionators.

4. The second characteristic is not apparent to the naked eye. Under a microscope you would be able to see that each catalyst particle has a large number of pores, and as a consequence, a tremendous surface area, especially in relation to the size of the particle.

5. The old alumina – based clay catalyst had the minerals necessary to promote the cracking reactions. Nowadays the catalysts are synthesized to exacting dimension and mineral content.

Ⅱ.1 – 5 BADAB 6 – 10 ABCAA

Ⅲ.1 – 5 CAAAA

Ⅳ.1. 催化剂和产物的混合物从提升管出来以后进入旋风分离器。旋风分离器是带动混合物高速旋转的一种机械装置。

2. 因为不同成分的原料与催化剂的反应速率不同,以这种进料方式可以处理不同的原料油。

3. 在老式的催化裂化装置中,沉降器称为反应器,因为大部分反应在其中进行。而在新式的催化裂化装置中,沉降器只是将催化剂从产物中分离出来。

4. 但是在一些提升管催化裂化装置中,一部分原料可以从提升管中部进料,从而减少接触时间。

5. 催化剂对反应的影响主要取决于催化剂与原料的接触程度,因此表面面积大在反应过程中是非常重要的。

Ⅴ.1. 为了解决这一现实经济问题,极具创造力的工程师们开发了一系列裂化技术,其中最著名的是催化裂化技术。

2. 第三个特性来自于技术的飞跃。早期白土催化剂含有促进裂化反应必需的天然矿物质,现在已经可以合成具有精确尺寸和矿物含量的催化剂了。催化剂的孔道可以精确设计制作,只允许特定尺寸的分子通过,一次一个,这样可以控制特定类型分子的反应方式。

4.2 催化裂化(2)

导语

催化裂化过程中一部分烃类形成积炭并覆盖在催化剂表面,待再生催化剂必须进入再生器再生。催化剂不断进行反应——再生的循环。裂化后的烃类产物离开反应器后输送至分馏塔,在分馏塔中切割成各种产品。催化裂化过程产生出与原油蒸馏不同的产品组分。

课文

催化裂化过程中总有一部分烃类形成积炭并覆盖在催化剂表面。催化剂表面被覆盖后,催化剂活性将下降(待生剂),且催化效率降低。为了消除积炭,待生剂通过自身重力进入再生器(图4.2)。在再生器中,待生剂与约 1100 °F 的热空气混合并发生下列反应:

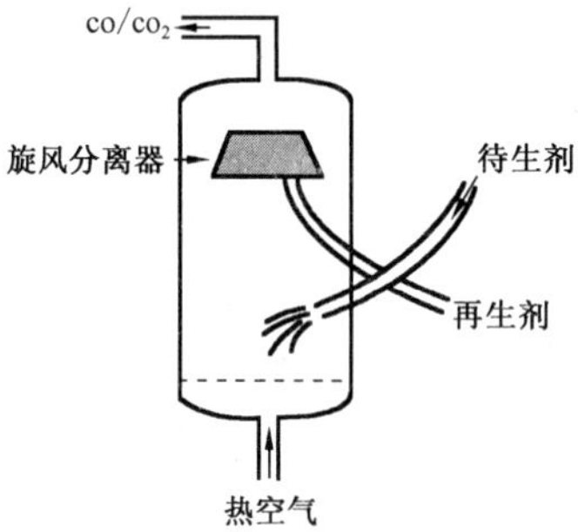

$$C + O_2 \longrightarrow CO + CO_2 (老式催化裂化装置)$$

$$C + O_2 \longrightarrow CO_2 (新式催化裂化装置)$$

图 4.2 再生器

积炭的氧化过程与锅炉中煤的燃烧类似,产生 CO_2 和大量的热,可能还有一部分 CO 生成。产生的热量以高温 CO 和 CO_2 形式存在并用于其他过程中,如用来驱动泵和透平机。在老式催化裂化装置中,CO 和 CO_2 流被输送到 CO 燃烧炉中,剩余的 CO 完全燃烧,然后排入大气。

再生器顶部有一个旋风分离器,它将催化剂与 CO/CO_2 气流分离。再生后的催化剂通过重力流出旋风分离器和再生器,然后与蒸汽混合后进入提升管反应器。从以上可以看出,催化剂不断进行着反应——再生的循环。

分馏系统

烃类裂化后的产物离开反应器后输送至分馏塔,在分馏塔中切割成各种产品(图4.3)。产品包括催化裂化气(C_4 及更轻的组分)、催化裂化汽油、催化裂化轻柴油、催化裂化重柴油（航煤），以及分馏塔底的油浆（称为循环油或回炼油）。循环油有多种加工途径，最常见的是与新鲜原料油混合后再次参与裂化反应。

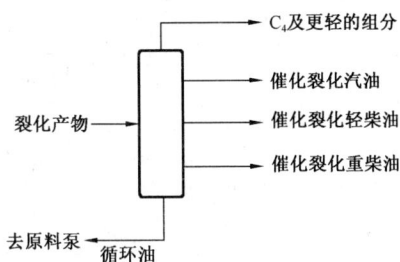

图 4.3　分馏塔

　　循环油经过多次循环，最终完全反应。这个过程有个名称——循环至消失。循环油中一些大分子仅进行循环而不再发生裂化。因此，少量循环油不断排出作为重质燃料油。

催化裂化生产的重柴油可以作为加氢裂化或者热裂化的原料，也用作残渣燃料油的调和组分。轻柴油是燃料油的优良调和组分，而催化裂化汽油是汽油的重要调和组分。

　　汽油和轻柴油的切割点有一定的范围。因季节变化，可以通过对切割点的控制来平衡汽油和馏分燃料油馏分。在冬季，取暖用馏分燃料油的需求增加，很多炼油厂采用燃料油产量最大化方案。通过调整催化裂化分馏塔，降低汽油馏分的终馏点，从而生产出更多的轻柴油。在夏季，通常采用汽油最大化方案，与冬季正好相反。

　　催化裂化分馏塔得到的产品组分与原油蒸馏得到的产品组分不同。催化裂化过程会产生烯烃，因此该过程生产的 C_4 及 C_4 以下的烃类不仅包括甲烷、乙烷、丙烷和丁烷，还有氢气、乙烯、丙烯和丁烯。由于这些额外的组分，催化裂化气需要输送到气体分离装置进行分离。而其他原油加工过程（如常压蒸馏、加氢处理、加氢裂化、催化重整等）产生的气体只含有饱和烃类，这些烃类的分离在饱和烃分离装置上进行。催化裂化所生产的异丁烷、丙烯和丁烯在烷基化过程中有重要用途，烷基化是生产汽油调和组分的过程。

　　催化裂化的其他产品在组成上也有所不同。在裂化过程中，很多大分子在苯环与侧链的连接处断裂，从而催化裂化产物中富含芳烃，而且苯环在分子结构中所占的比例也有增大的趋势。这对提高汽油的辛烷值有利，而对航空煤油和柴油却不利。

　　整个催化裂化装置如图 4.4 所示。从图中可以看到有两个循环流。在图的左边部分，催化剂经历反应、再生、再反应这样的循环；在图的右边部分，原料进入装置，产物离开装置，而循环油里的一些组分不断循环。

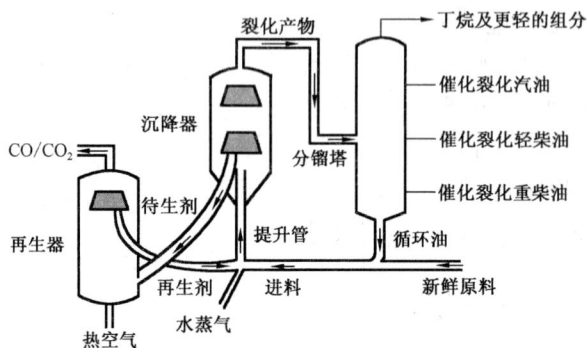

图 4.4　催化裂化装置

表 4.1　催化裂化原料和产品组成

项目	组成, %（体积分数）
原料:常压渣油	40.0
减压馏分油	60.0
循环油	(10.0)*
合计	100
产品:焦炭	8.0
C_4 及各轻的产品	35.0
汽油	55.0
轻柴油	12.0
重柴油	8.0
循环油	(10.0)*
合计	118.0

＊原料和产品不考虑循环油。

　　由于循环油最终会转化为产品而消失,因此在催化裂化装置简图中并不显示循环油的流程。然而另一个重要现象——体积的增加出现了。从产品收率分布表(表 4.1)可以看出,产品的总体积是原料的 118%,这和循环油无关,而是由于原料和产品的密度不同造成的。如果以质量收率代替体积收率,那么总的产品收率仍是 100%。然而美国大部分石油产品的销售都是以加仑为计量单位的,因而炼厂均按体积衡量收率。催化裂化过程中会引起密度的变化,结果造成体积的增加。体积的增加在统计上带来麻烦,而在炼油厂却可以作为获利的手段,"稀释每一桶油"。

🔲 **Key to Exercises**

　　Ⅰ.1. To remove the carbon,the spent catalyst flows by gravity to a vessel called a regenerator.

2. On the left side, the catalyst goes through the reaction, is regenerated and gets charged back to the reaction again. On the right side, hydrocarbon comes in and goes out, but the cycle oil provides continuous circulation of at least some of the hydrocarbon components.

3. As the winter heating oil season arrives, many refineries go into a max distillate mode. They make adjustments to the CCU fractionator to lower the end point of the cat cracked gasoline to push more volume into the cat light gas oil.

4. The C_4 and lighter stream not only contain methane, ethane, propane, and butanes but also hydrogen, ethylene, propylene, and butylenes.

5. A variety of things can be done with the cycle oil but the most popular is to mix it with the fresh cat feed and run it through the reaction again.

Ⅱ. 1 – 5　ABACD　6 – 8　BCA

Ⅲ. 1 – 5　BAAAB

Ⅳ. 1. 产生的热量以高温 CO 和 CO_2 形式存在并用在其他过程中,如用来驱动泵和透平机。

2. 从以上可以看出催化剂不断进行着反应—再生的循环。

3. 与此同时,烃类裂化后的产物离开反应器以后输送到分馏塔中,在分馏塔中切割成各种产品。

4. 汽油和轻柴油的切割点有一定的范围。

5. 催化裂化分馏塔得到的产品组分与原油蒸馏得到的产品组分不同。

Ⅴ. 1. 催化裂化生产的重柴油可以作为加氢裂化或者热裂化的原料,也用作残渣燃料油的调和组分。轻柴油是燃料油的优良调和组分,而催化裂化汽油是汽油的重要调和组分。

2. 整个催化裂化装置如图所示。从图中可以看到有两个循环流。在图的左边部分,催化剂经历反应、再生、再反应这样的循环;在图的右边部分,原料进入装置,产物离开装置,而循环油里的一些组分不断循环。

4.3　催化重整

导语

催化重整是一种非常重要的工艺,可用来把低辛烷值的石脑油炼制成作为高辛烷值汽油调和组分的重整油,不仅提高了汽油的质量,还提高了汽油的产量。但是催化重整装置生产的汽油中含有苯,引起了公众对环境的关注,同时也引起了炼油工作者和环境保护者之间的辩论。

课文

在炼油工业中,催化重整工艺比其他所有工艺更饱受争议。开始是为了解决市场需求,到后来则演变成了一场炼油工作者和环境保护者之间的辩论。

发展历史

从催化重整技术成为提高汽油产量和质量的手段开始,就注定了其曲折的发展历程。在 20 世纪上半叶,汽油的需求量就以比燃料油快两倍的速率增长。而汽车生产商所设计的汽车对汽油规格要求也不断增加。美国修建了 4 车道和 6 车道的高速公路贯穿了整个大陆,消费者出行不再乘坐福特 T 型车,而是乘坐宽大舒适的客车,而这些车都是以便宜的高辛烷值汽油为燃料的。

催化重整作为一项精炼技术,需要寻找到合适的组分作为原料,人们发现从煤油中分离出的重石脑油最合适。催化重整过程能提高这些石脑油的质量,使其辛烷值从 30 ~ 40 提高到 90 以上。同时,催化重整技术的应用不仅提高了汽油的生产量,还提高了汽油的质量。

黄金(大发展)时期。1949 年美国环球油品公司开发了现代化的催化剂和工业设备。他们正好赶上了第二次世界大战后 25 年的黄金发展时期,在那期间,美国保持了 7% 的年均经济增长率。那时,炼油厂商一切以辛烷值为主,不断地广告宣传更高辛烷值的汽油,辛烷值 100 的汽油就是碳氢化合物的"圣杯"。而为了实现广告中的承诺,炼油厂商必须建立新的炼油厂,改进催化重整装置,进一步提高汽油质量。

环境问题。在 20 世纪 70 年代,加强环境保护提上了公众的议程,政府开始要求炼油厂减少汽油中铅的添加量。经过 10 年的努力,工程师们发明了一项技术,在汽油里加入极少量的四乙基铅,可以大幅提高汽油的辛烷值。在 20 世纪 70 年代中期,虽然政府早就了解铅的毒害,但还只是制定了一个时间表,在未来的 10 年里逐步淘汰铅的使用。炼油工程师重新设计了催化重整工艺,消除了旧工艺中的不足之处。同时开发了一种新的催化剂,用于现行催化重整工艺,以增加汽油中芳烃的含量,这是高辛烷值汽油中重要的调和组分。

在 20 世纪 80 年代,随着汽油中铅的淘汰,本德定律(新污染不除旧污染不去)应验,汽油中含苯所造成的影响又变成了一个大问题。众所周知,苯是一种致癌物质,虽然使用催化重整装置生产的汽油中含有极少量的苯,但这已经引起了公众的关注,最后逐步制定法规,减少苯的含量。炼油工程师再一次研究改变原料,提高催化重整装置催化剂的性能,用其他无害产物代替苯,消除污染。经过努力,芳烃总的含量得到了控制。

主力副产品。催化重整技术的发展使人们逐渐意识到催化重整的一种副产品,即氢气,已经成为炼油厂的主力。在加氢处理过程中,氢气用来帮助脱除各种较脏馏分中的硫和其他污染物,以及改变其他一些分子的结构。氢气更是加氢裂化装置最重要的原料。现在当催化重整装置因为维修或者紧急情况而关闭时,炼油厂的主要部门都将受到氢气短缺的影响而不得不关闭。

在催化重整中发生的有益反应主要如下:
- 正构烷烃转变为异构烷烃。
- 正构烷烃转变为环烷烃,并释放出氢气。
- 环烷烃转变为芳烃,并释放出氢气。

从对辛烷值的影响角度来看,以下一些不利的反应也同时进行:
- 一些正构烷烃和环烷烃发生裂化反应,形成了丁烷和更轻的气体。
- 一些环烷烃和芳烃的侧链断裂,形成了丁烷和更轻的气体。

其中,最重要的是正构烷烃和环烷烃转变为异构烷烃和芳烃。

催化重整装置

人们可能认为在催化重整中是用到一些非常独特的设备,才使得这些复杂的反应在重整过程中发生。但正好相反,所需要的仅仅是一种由铝、硅、铂和钯构成的具有独特功能的催化剂。在生产中,铂的用量不少(在一套装置中就需要几百万美元的铂),所以人们开始关注对铂的研究。金属铂和钯在催化剂中是最重要的组分,能够促使正构烷烃卷曲成环,脱去多余的氢,拨开以前没有的侧链。

有很多方法可以使碳氢化合物接触到催化剂。这里提到的方法称为固定床,有一个或者多个反应器,碳氢化合物逐步穿过容器中的催化剂,如图4.5所示。

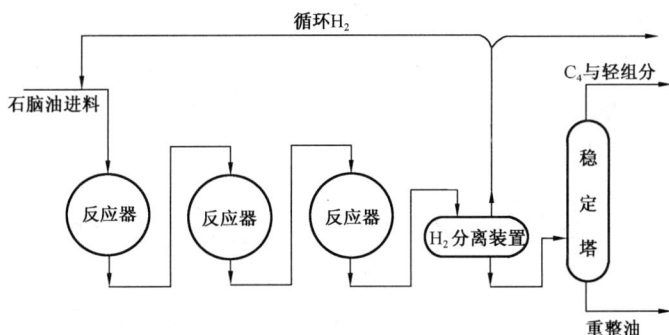

图 4.5　催化重整工艺流程

原料石脑油经过加压、加热后,在第一个反应器中通过催化剂床层进行反应,最后从反应器的底部流出。这个过程在接下来的两个反应器中重复进行。所生产的产品经过冷却器,大部分液化。液化的目的是为了分离富氢气流以便循环使用。

同时,在反应器的外面,一部分氢气重新循环进入原料,其他部分则输送到气体处理装置。从反应器底部分离出来的液态产物输送到稳定塔,即脱丁烷塔。从稳定塔底部出来的产物称为重整油,丁烷和轻组分从顶部输送到气体处理装置。

🔲 小结

催化重整是一种非常重要的工艺,可用来把低辛烷值的石脑油炼制成作为高辛烷值汽油调和组分的重整油。正构烷烃转变为异构烷烃,环烷烃转变为芳烃,都是有效提高辛烷值的途径。然而重整油的辛烷值越高,最终的产物收率也就越低,而生产出的轻组分就越多。

🔲 Key to Exercises

Ⅰ. 1. Cat reforming, has provided more controversy in the refining business than all the other units combined.

2. In 1949, Universal Oil Products Company introduced the present day catalyst and plant design.

3. In the 1970s, enough concern about the environment reached the public agenda that governments began requiring refiners to reduce the amount of lead put in gasoline.

4. In the 1980s the benzene content of gasoline became a big issue.

5. Hydrogen emerged as an essential workhorse in the refinery. Hydrogen was increasingly used in hydrotreaters to help remove sulfur and other contaminants from various dirty streams and change the structure of some others.

Ⅱ. 1 – 5　ACDBA　6 – 10　BCACA

Ⅲ. 1 – 5　ACBCA

Ⅳ. 1. 美国修建了 4 车道和 6 车道的高速公路贯穿了整个大陆,消费者出行不再乘坐福特 T 型车,而是乘坐宽大舒适的客车。

2. 为了实现广告中的承诺,炼油厂商必须建立新的炼油厂,改进催化重整装置,进一步提高汽油质量。

3. 经过 10 年的努力,工程师们发明了一项技术,在汽油里加入极少量的四乙基铅可以大幅地提高汽油的辛烷值。

4. 所生产的产品经过冷却器,大部分液化。液化的目的是为了分离富氢气流以便循环使用。

5. 人们可能认为利用一些非常独特的设备可使这些复杂的反应在重整过程中发生。

V.1. 同时,在反应器的外面,一部分氢气重新循环进入原料,其他部分则输送到气体处理装置。从反应器底部分离出来的液态产物输送到稳定塔,即脱丁烷塔。从稳定塔底部出来的产物称为重整油,丁烷和轻组分从顶部输送到气体处理装置。

2. 催化重整是一种非常重要的工艺,可用来把低辛烷值的石脑油炼制成作为高辛烷值汽油调和组分的重整油。正构烷烃转变为异构烷烃,环烷烃转变为芳烃,都是有效提高辛烷值的途径。然而随着重整油的辛烷值越高,最终的产物收率也会越低,而得到的轻组分也越多。

4.4 加氢裂化

📖 导语

加氢裂化工艺的作用比热裂化、催化裂化和催化重整都大。加氢裂化可以以轻瓦斯油或重瓦斯油为原料生产质量更高的汽油,而且加氢裂化过程没有塔底残渣产生,所有的产物都是轻油。加氢裂化可灵活调整炼厂生产,冬天利用加氢裂化最大化地生产柴油和燃料油,夏天最大化地生产汽油和航空煤油,同时能提高质量。另外,加氢裂化可以为烷基化装置提供异丁烷或者为重整装置提供石脑油,并且可以通过催化重整过程生产氢气。

📖 课文

加氢裂化工艺比热裂化、催化裂化和催化重整出现得晚,但是该工艺的作用却比上述任何一个都大。加氢裂化工艺的引入可以解决许多令炼油厂头痛的问题,如不同月份或季节之间由于产品市场变化引发的问题。加氢裂化可以以轻瓦斯油或重瓦斯油为原料生产汽油,其质量比通过热裂化工艺生产的汽油要好。加氢裂化也可以利用重瓦斯油生产轻的馏分油(航空煤油和柴油)。加氢裂化还可以生产相对大量的异丁烷,可以为烷基化装置提供原料。最重要的是,加氢裂化过程没有塔底遗留的残渣(焦炭、沥青或残油)产生,所有的产物都是轻油。冬天利用加氢裂化最大化地生产柴油和燃料油,夏天最大化地生产汽油和航空煤油。加氢裂化可灵活调整炼厂生产。

工艺流程

加氢裂化是在氢气存在条件下的催化裂化过程,工艺简单。通过氢气、催化剂和适当操作条件可以加工各种原料,如轻瓦斯油、直馏渣油,催化裂化或热裂化循环油。加氢裂化工艺分为多个阶段,每个阶段都要给组分升级——如重质油到中间馏分油,中间馏分油到汽油组分。

为什么不是每个炼油厂都拥有加氢裂化装置?尽管目前大约有12种加氢裂化工艺很受欢迎,但是这些装置的建造和操作成本非常昂贵。下面介绍比较常用的工艺。

设备和反应

与重整催化剂相比,加氢裂化催化剂的成本较低,一般是钴、钼或者是镍加上铝的硫化物。与催化裂化不同,但是与催化重整相似,加氢裂化使用的催化剂在固定床上反应。就像催化重整过程,加氢裂化过程是在两个反应器进行,如图4.6所示。

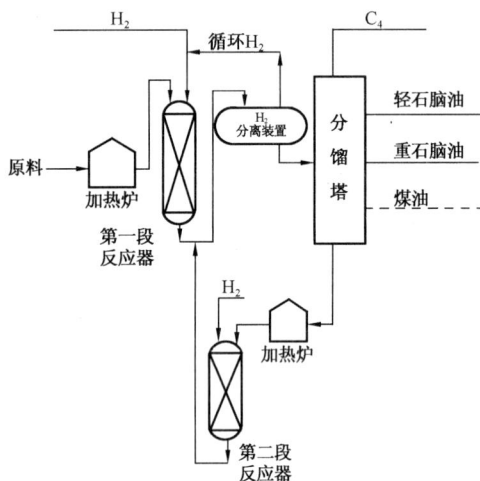

图4.6 加氢裂化工艺流程

原料(重瓦斯油)与氢气混合后加热到 550 ~ 750 ℉,加压到 1200 ~ 2000psi,输入至一段反应器中。通过催化剂床层后,40% ~ 50% 的原料裂化成为汽油成分(终馏点低于 400℉)。

氢气和催化剂在多方面互补。催化剂引起长链分子裂化,使得芳香化合物的环打开。这两种反应都需要加热以保持反应进行,是一个吸热的过程。当裂化反应发生时,过量的氢气使分子饱和,这个过程放出热量,称为加氢作用,是一个放热过程。这样,加氢过程放出的热量可以保持裂化过程的进行以及环的打开。

反应产物离开一段反应器后,大部分冷却液化,进入氢气分离器。在氢气分离器中,氢气分离出来后继续作为原料循环使用,液体进入分馏塔中。无论想生产什么产品(汽油、航空煤油以及柴油),分馏塔都可以从一段反应器中的液态产物中将它们分割出来。塔底产物可以继续在二段反应器中反应。换句话说,煤油和轻柴油系列可以作为分馏塔侧线产品获得,如果以汽油为产品目标,就会将这些组分留在塔底继续转化成汽油。

分馏塔塔底产物再次与氢气混合输送到二段反应器。因为这些原料已经在一段反应器中经过了加氢、裂化和重整反应,二段反应器的操作更加苛刻(高温高压)。像一段的一样,二段产品从氢气中分离出去后输送到分馏塔中。

有些加氢裂化过程分为三个阶段,要么在一个反应器中对不同催化剂进行堆叠,要么是设置三个反应器。在每一个阶段都有不同的反应发生。在第一阶段,催化剂把含有硫、氮和污染物等交织在一起的复杂分子的环打开,氢气则与硫和氮分别形成 H_2S 和 NH_4。同时,氢气与打开的双环反应,形成简单的较轻化合物。

在第二阶段,运行条件更苛刻,催化剂使更多重组分的环打开,进行裂化反应,形成轻质产品。在第三阶段,芳烃和烯烃饱和生成环烷烃、烷烃,尤其是异构烷烃。

加氢裂化是在相当苛刻的条件(2000psi 和 750 ℉)下进行的。可以设想一下加氢裂化的反应器。使用专业钢生产的反应器器壁有 6 英寸厚。最担心的是裂化过程失控,因为所有的过程都是吸热反应,快速升温可以促进裂化反应的进行,只有发生加氢反应后才能放出更多的热。因此,在大部分的加氢裂化装置里安装了精密的控温系统来控制这个过程。

重油加氢裂化

有些重油加氢裂化装置是用来加工常压渣油或者减压渣油的,其操作与加氢处理过程类似。90% 以上的产物都是渣油燃料。操作的目的是通过氢气和硫化物的催化反应形成 H_2S,脱除硫。硫含量为 4% 左右的渣油经过处理,可以将硫含量降至低于 0.3%。

📮 小结

随着加氢裂化在炼油厂的出现,显然要求操作流程绝对一体化。加氢裂化工艺是炼油厂的重点技术,因为该工艺可以调节炼油厂汽油、馏分油、航空煤油的产量,同时能提高它们的质量。加氢裂化过程取决于进料速率、催化裂化、焦化或热裂化等运行条件。另外,加氢裂化可以为烷基化装置提供异丁烷或者为重整装置提供石脑油,并且可以通过催化重整过程生产氢气。

🔲 Key to Exercises

Ⅰ. 1. Hydrocracking is a process of more recent vintage than thermal or cat cracking or cat reforming, but it was designed to accomplish more of what each of the earlier processes do.

2. Refiners use hydrocrackers to move from max diesel and distillate fuels in the winter to max gasoline and maybe even jet fuel in the summer.

3. Hydrocracking can simultaneously improve the quality of both the gasoline blending and the distillate fuel blending pools.

4. The hydrogen and the catalyst are complementary in several ways. First, tile catalyst causes long chain molecules to crack and the rings in aromatic compounds to open. Both these reactions need heat to keep them going. They are both an endothermic process. On the other hand, as the cracking takes place the excess hydrogen floating around saturates (fills out) the molecules, a process that gives off heat. This process, called hydrogenation, is exothermic. Thus, hydrogenation gives off the heat necessary to keep the cracking and ring opening going.

5. A critical worry is the possibility of runaway cracking. Since the overall process is endothermic, the temperature can rise rapidly, accelerating the cracking rates dangerously, which only gives off more heat as hydrogenation then takes place.

Ⅱ. 1 - 5　BDACA　6 - 10　ABABA

Ⅲ. 1 - 5　AAAAD

Ⅳ. 1. 它们的质量比通过热裂化工艺生产的汽油要好。

2. 通过氢气、催化剂和适当操作条件可以加工各种原料,如轻瓦斯油、直馏渣油以及催化裂化或热裂化循环油。

3. 尽管目前大约有 12 种加氢裂化工艺很受欢迎,但是这些装置的建造和操作成本非常昂贵。

4. 这两种反应都需要加热以保持反应进行,是一个吸热的过程。

5. 分馏塔塔底产物再次与氢气混合输送到二段反应器。

Ⅴ. 1. 反应产物离开一段反应器后,大部分冷却液化,进入氢气分离器。

在氢气分离器中,氢气分离出来后继续作为原料循环使用,液体进入分馏塔中。无论想生产什么产品,分馏塔都可以从一段反应器中的液态产物中将它们分割出来。塔底产物可以继续在二段反应器中反应。

2. 在第二阶段,运行条件更苛刻,催化剂使更多重组分的环打开,进行裂化反应形成轻质产品。在第三阶段,芳烃和烯烃饱和生成环烷烃、烷烃,尤其是异构烷烃。

4.5 烷基化

导语

烷基化是指利用丙烯、丁烯与异丁烷合成异构烷烃的反应,这种异构烷烃称为烷基化油。与催化裂化工艺相反,烷基化工艺主要是以小分子合成大分子。烷基化装置主要由七部分组成。烷基化装置必须对一系列重要参数进行设置,抑制过多副反应发生,以免降低烷基化油的质量。烷基化油提高了汽油的环境质量,是理想的调和组分,因其蒸气压低,不含硫,不含烯烃,不含苯,并具有较高的辛烷值。

课文

催化裂化工艺发明后,工程师们便开始解决该过程产生的所有轻质烃问题,即想要汽油产量最大化,但是工艺过程中大量挥发性强的丁烯和丙烯又难以溶解在汽油中。由此,烷基化工艺得以发展。与催化裂化工艺相反,烷基化工艺主要是以小分子合成大分子。

化学反应

对于化学家而言,烷基化包括了一些小分子合成反应。对于炼化企业员工而言,烷基化是指利用丙烯、丁烯与异丁烷合成异构烷烃的反应,这种异构烷烃称为烷基化油(图4.7)。

由于在烷基化反应过程中分子存在很大程度的收缩,在石油炼制过程中,烷基化具有与裂化相反的体积效应。当使用丙烯为原料时,1桶丙烯和1.6桶异丁烷反应可生成2.1桶产品;1桶丁烯和1.21桶异丁烷反应可生产1.8桶产品。与裂化反应一样,反应过程中质量守恒,不存在质量损失,改变的只是密度和体积。

工艺流程

丙烯和丁烯的化学反应活性很高,只需将异丁烷与烯烃的混合物置于

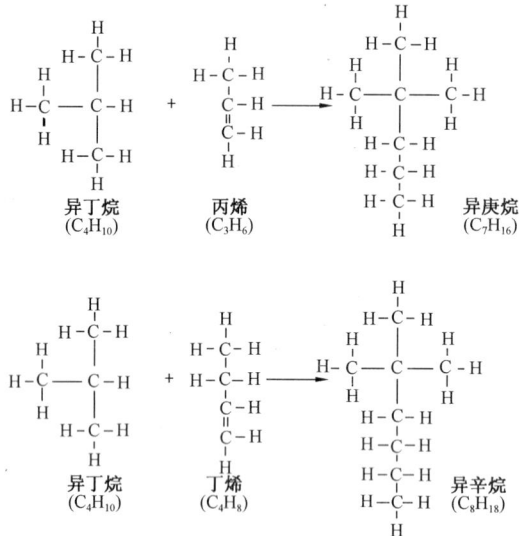

图 4.7　丙烯和丁烯的烷基化

高压环境下便可发生反应。然而进行这种烷基化操作的设备成本太高。与其他反应过程类似,催化剂的开发可以促进烷基化反应,简化反应流程。烷基化装置使用硫酸和氢氟酸作为催化剂。烷基化催化剂为液体,而催化裂化催化剂为固体,但两者的反应过程基本一致。使用氢氟酸作为催化剂的装置称为氢氟酸烷基化装置,其他的装置则称为硫酸烷基化装置。因为氢氟酸和硫酸都具有很强的腐蚀性,能腐蚀所有的容器和管线,除非有防腐内衬,所以使用氢氟酸和硫酸作为催化剂必须关注安全性。尤其是氢氟酸,一旦发生泄漏,便会漂浮在云里,传播到很远的地方,严重影响炼化企业周围的环境。而硫酸发生泄漏则会呈液滴状,会快速滴落到地面,但不会对周边工作人员造成很大影响。

　　硫酸烷基化装置适用于丁烯烷基化,而氢氟酸烷基化装置则更适用于丙烯烷基化,前者应用更为广泛,故本文只介绍硫酸烷基化装置。事实上,氢氟酸烷基化工艺也大致相似。

　　烷基化装置主要由七部分组成:冷却器、反应器、酸分离器、碱洗罐以及3个分馏塔,如图4.8所示。

　　(1)冷却器。硫酸催化烷基化反应的适宜温度约为40 °F。烯烃(来自裂化装置的丙烷或丙烯、丁烷或丁烯)和异丁烷同时输送至冷却器和硫酸混合,并且保持足够压力使混合物呈液态。有些情况下,冷却器和反应器合为一体。

　　(2)反应器。烷基化反应时间较长,因此,反应混合物输至一组大反应

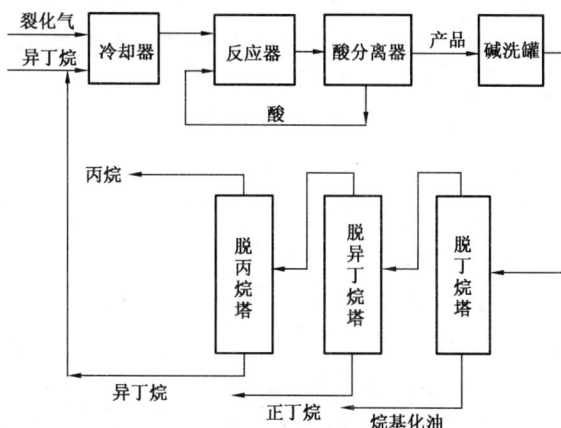

图 4.8　烷基化装置

器中。反应器的体积大到足以保证所有的反应混合物在特定时间内完全反应。反应物停留时间为 15～20 分钟。当反应物流经过反应器时,搅拌器对其进行搅拌,使得烯烃和异丁烷与硫酸催化剂充分混合,确保反应进行。

（3）酸分离器。从反应器中流出的反应混合物流入酸分离器中静置,酸液和液态烃会像油水一样分离。液态烃从上层抽出,而酸液从下层放出,酸液随后又循环回进料处。酸分离器又称为酸沉降器。

（4）碱洗罐。从酸分离器中抽出的液态烃还含有一些残酸,必须在碱洗罐中用氢氧化钠进行脱酸处理。氢氧化钠对液态烃起着中和酸的作用,正如消食片对消化不良的胃的作用一样。残留的即为待分离的烃类混合液。

（5）分馏塔。3 个标准的分馏塔用于分离烷基化油与饱和气体,没有反应的异丁烷重新返回到进料处。

工艺参数

烷基化副反应在很大程度上会降低烷基化油的质量,如产品辛烷值低,色泽度差,蒸气压过大。因此,烷基化装置必须对一系列重要参数进行设置,以抑制过多副反应的发生。

（1）反应温度。温度过低会使硫酸变得黏稠,从而导致反应物混合不均匀,不能充分反应;而过高的温度除了会生成异庚烷和异辛烷之后,还会生成其他的一些副产物,从而降低烷基化油的质量。

（2）酸强度。由于循环使用,酸液不可避免地受到液态烃和焦油中的水分稀释。当酸强度从 99% 降低到 89% 时,就必须抽出送回酸储罐脱水。

（3）异丁烷浓度。异丁烷比例越大,产品质量越高。反应器中通常设置异丁烷循环系统。异丁烷与烯烃的比例设置在 5∶1～15∶1 之间。一般通过反应器中的隔间来控制异丁烷浓度。

(4)烯烃反应速率。初级烯烃原料在反应器中的停留时间对烷基化油的质量有很大影响。

🔲 小结

近年来,为提高汽油的环境质量,烷基化起着极其重要的作用。在汽油调和方面,烷基化油是理想的调和组分,因其蒸气压低,不含硫,不含烯烃,不含苯,并具有较高的辛烷值,而所有这些都是用炼油厂里的边角料诸如丙烯、丁烯和异丁烷合成的。

在炼厂流程图中,烷基化过程可以用一个方框图表示,进料为丙烯、丁烯、丙烷和正丁烷,出料为烷基化油、丙烷和正丁烷。图4.9描述的是到目前为止介绍过的炼油厂所有生产装置,包括烷基化装置。

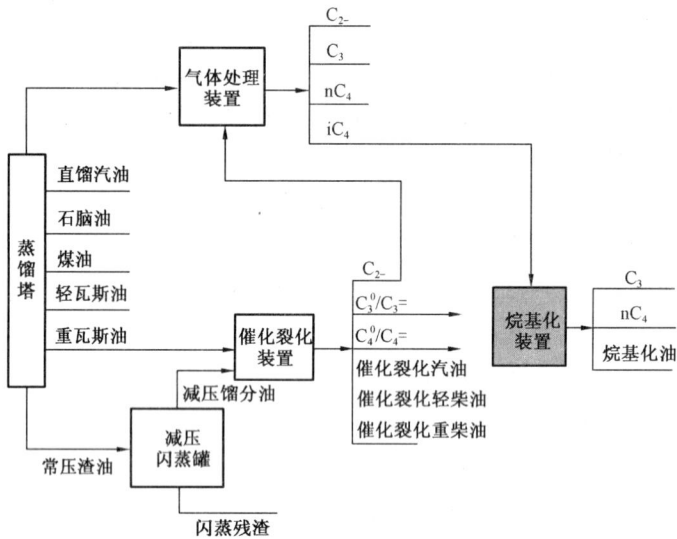

图4.9 炼油厂流程图

🔲 **Key to Exercises**

Ⅰ.1. After the engineers were so clever about the invention of cat cracking, they attacked the problem of all the light ends the process created.

2. Propylene and butylenes are hyper enough that the chemical reaction could be made to take place by just subjecting the isobutane and olefins, to high pressures. However the equipment would be very expensive to handle this route

to alkylation.

3. The alky plant consists of seven main parts: the chiller, the reactors, the acid separator, the caustic wash, and three distilling columns.

4. The alky plant manager has to watch a number of key variables to keep too many side reactions from occurring that could cause the quality of the alkydate to deteriorate, as evidenced by such things as lower octane umber, poor color, and high vapor pressure.

5. Alkylate has emerged as a hero in the past few years as refiners struggle to improve the environmental qualities of gasoline.

Ⅱ.1－5　BCDAB　6－10　ABDAC

Ⅲ.1－5　BCBBB

Ⅳ.1. 对于化学家而言,烷基化包括了一些小分子合成反应。

·2. 由于在烷基化反应过程中分子存在很大程度收缩,在石油炼制过程中,烷基化具有与裂化相反的体积效应。

3. 然而进行这种烷基化的操作成本太高。

4. 而硫酸发生泄漏则会呈液滴状,会快速滴落到地面,不过不会对周边工作人员造成很大影响。

5. 酸液随后又循环回进料处。酸分离器又称为酸沉降器。

Ⅴ.1. 硫酸烷基化装置适用于丁烯烷基化,而氢氟酸烷基化装置则更适用于丙烯烷基化,前者应用更为广泛,故本文只介绍硫酸烷基化装置。事实上,氢氟酸烷基化工艺也大致相似。

2. 近年来,在炼油厂商为提高汽油的环境质量努力的过程中烷基化起着极其重要的作用。在汽油调和方面,烷基化油是理想的调和组分,因其蒸气压低,不含硫,不含烯烃,不含苯,并具有较高辛烷值。而所有这些都是用炼油厂里的边角料诸如丙烯、丁烯和异丁烷合成的。

第5章 石油炼制产品和化工产品

5.1 石油炼制产品（1）

🔲 导语

19世纪60年代，现代炼油技术刚刚起步时，从原油提炼出的主要产品只有照明煤油和作润滑油的渣油。然而现在大部分原油最终都成为某一燃料油。从原油炼制的燃料油可分为两大类：与空气雾化后燃爆，用以提供原动力的燃料油；直接燃烧用以取暖和照明，或转换为像电力这样二次能源的燃油。前一种燃料油包括航空油（航空汽油、航空涡轮汽油、航空涡轮煤油）、工业及民用液化气（LPG）、车用汽油、柴油和炼厂气，后一种包括普通煤油及各种级别的燃料油。

🔲 课文

汽油

汽油是以碳氢化合物为主的复杂混合物，沸点低于200℃（392℉）。汽油的组成成分差别很大，即使含有相同辛烷值的汽油，其成分也不尽相同。由于不同汽油的组分不同，因此有必要进行汽油混合。混合的物理过程很简单，但是决定每种成分在混合物中的比例比较困难。操作时需要把汽油的所有成分同时泵入通往汽油储库的管道。油泵一定要设置成可以自动按比例输送每一种组分的模式。能实现预期混合物的仪器也很复杂。

汽油（实际是产品）混兑的目的是调配现有组分，以满足产品的需要和规格。在混兑过程中，把其他装置中的产品流收集到一起，混兑成预期的产品。例如，加氢精制装置、重整装置、叠合装置和烷基化装置中的产品流经混合生产出符合规格的汽油。

北方冬季气温极低，因而汽油里可以有一定量的丁烷，以利于汽油蒸发。沸程越窄，飞机引擎的感应系统越复杂，汽化燃料的分布就越均匀。然而新配方油可能会增加油耗，这时不混合更为可取。生产低硫汽油（和柴油）是保证空气清新和减少因交通而引起的空气污染的必要措施。炼油厂正在使用加氢技术，即原料（无论是馏分物还是渣油）用氢处理来减小硫含量，从而使加工更简便。

煤油

煤油,也称为"kerosine",是石油在 205～260℃的温度范围内产生的直馏(蒸馏)馏分。煤油在汽车时代之前曾是主要的炼制产品,而如今的煤油也许只能定义为汽油之后的其他石油产品之一。在炼油初期,一些原油含有高质量的煤油馏分,但是其他原油,如含有较高沥青馏分的原油必须彻底精炼,去除芳烃及含硫化合物后才能得到合格的煤油馏分。煤油主要成分为含 12～15 个碳原子的烃分子。低含量的芳烃与不饱和烃可使煤油燃烧时排烟量保持最低值。虽然一些芳烃可能会在煤油的沸程内产生,但是可以通过提取的方式取出其中大量的芳烃。煤油中的总硫含量会因油品类型和用途而有很大不同。硫含量问题之所以非常重要,是因为当煤油燃烧产生硫氧化物时,会带来环境问题。煤油的颜色不是很重要,但由于污染或老化,产品颜色会比正常颜色深。实际上,煤油颜色比指定颜色深可能会令使用者不满。煤油作为燃料使用,所以不能含有芳烃与不饱和烃。煤油的理想成分是饱和烃。柴油机燃油、航空燃油、煤油(炊事用煤油)、1 号燃料油、2 号燃料油都是常见的煤油馏分产生的蒸馏产品。航空燃油中的某个等级产品用的是重质石脑油馏分,但煤油馏分却是航空燃油中更常见、更重的成分,煤油馏分还有少量作为燃料油或 1 号燃油出售。一些燃油(通常是 2 号燃油)和柴油燃料类似,有时能互相替代。

燃料油

燃料油有多种分类方法,通常分为两大类:馏分燃料油和残渣燃料油。馏分燃料油在蒸馏过程中蒸发浓缩,有明确的沸程,且不包括高沸点油或沥青组分。

原油蒸馏物加氢裂化得到的含有一定量残渣的燃料油称为残渣燃料油。然而馏分燃料油和残渣燃料油的这两种定义的界限正逐渐模糊,因为燃料油既可以是馏分油又可以是残渣油,以满足特殊用途。目前国内燃料油、柴油和重燃油定义更多地表述的是燃料油的用途。国内主要生产家用馏分油,包括煤油、炉用油和锅炉燃料油,柴油也属于馏分油。而残渣油已经成功应用于船用柴油机中,馏分油和残渣油的混合油品也已经用于机车柴油。

重燃油包含多种油品,从馏分油到残渣油,但在使用前必须加热到260℃或更高温度。通常情况下,为了满足特殊要求,重燃油是由混合了馏分油的残渣油组成的。重燃油包括各种工业用油,当给船作燃料时,称为船用油。炉用油是重质油中的直馏(蒸馏)组分,而其他的燃料油通常调和两种或更多组分。能够调和入燃料油的直馏组分有重石脑油、轻瓦斯油和重瓦斯油以及渣油。来自催化裂化、裂解焦油和催化裂化分馏塔塔底的轻、重瓦斯油等馏分可混入不同的燃料油,以满足性能要求。重燃油常含有渣油组分,渣油与瓦斯油和分馏塔塔底组分混合达到一定黏度。工业上,有些时候需要使火焰或烟道气接触产品(例如制陶业、制玻璃业、热处理和平炉),此时燃料油必须经调和以降低硫含量;这些含硫低的燃料油更受欢迎。燃料油加工过去曾经使用原油中提取出预期产品后的剩余物。现在的燃料油生产是通过选择与混合不同原油组分来满足特定需求。用于加热的燃料油分1号到6号6个等级,包括轻馏分油、中馏分油、重馏分油、馏分油和残渣的混合组分以及渣油。

润滑油

润滑油以高沸点(>400℃)与高黏度区别于其他石油馏分。润滑油可以根据用途分为许多种类,但主要分为两大类:用于间歇性润滑的油,如车用机油和航空油;用于连续润滑的油,如汽轮机油。用于间歇润滑的润滑油黏度不应受温度变化影响。为了除去使用中累积的外来杂质,此类润滑油必须频繁地更换。相比之下,用于连续润滑的润滑油在长时间使用过程中不能更新,因此它的稳定性能更加重要。用于连续润滑的润滑油必须特别稳定,因为要求发动机持续保持在恒温下运行。

Key to Exercises

I. 1. The objective of gasoline (in fact, product) blending is to allocate, the available blending components in such a way as to meet product demands and specifications.

2. The production of low sulfur gasoline (and diesel fuel) is a necessary step to assure clean air and to reduce air pollution from the transportation sector.

3. In the early days of petroleum refining, some crude oils contained kerosene fractions of high quality, but other crude oils, such as those having a high proportion of asphaltic materials, had to be thoroughly refined to remove aromatics and sulfur compounds before a satisfactory kerosene fraction could be obtained.

4. Fuel oil is classified in several ways, but generally into two main types: distillate fuel oil and residual fuel oil.

5. Lubricating oil is distinguished from other petroleum fractions by the high (>400℃) boiling point as well as their high viscosity.

Ⅱ.1 – 5　CCBAC　6 – 10　CDAAC

Ⅲ.1 – 5　BDDAC

Ⅳ.1. 汽油是以碳氢化合物为主的复杂混合物,沸点低于 200℃(392℉)。

2. 生产低硫汽油(和柴油)是保证空气清新和减少因交通引起空气污染的必要措施。

3. 然而,新配方油可能会增加油耗,这时不混合更为可取。

4. 煤油在汽车时代之前曾是主要的炼制产品,而如今的煤油也许只能定义为汽油之后的其他石油产品之一。

5. 硫含量问题之所以非常重要,是因为当煤油燃烧产生硫氧化物时,会带来环境问题。

Ⅴ.1. 汽油(实际是产品)混兑的目的是调配现有组分,以满足产品的需要和规格。在混兑过程中,把其他装置中的产品流收集到一起,混兑成预期的产品。例如,加氢精制装置、重整装置、叠合装置和烷基化装置中的产品流经混合可生产出符合规格的汽油。

2. 燃料油有多种分类方法,通常分为两大类:馏分燃料油和残渣燃料油。馏分燃料油在蒸馏过程中蒸发浓缩,有明确的沸程,且不包括高沸点油或沥青组分。

5.2　石油炼制产品（2）

导语

现在大约 88% 的原油最终都成为某一燃料油。在余下的 12% 原油中,只有一半略多一点炼制成石化半成品。这些半成品是制成合成材料(纤维、橡胶、塑料等)、化肥、农药,甚至动物饲料蛋白的原料。平均每桶原油大约有 5% 用于生产各种润滑油、润滑脂、石蜡、溶剂以及铺路和防水用的沥青。

课文

蜡

石油蜡一般分为两种:馏分油中的石蜡和残渣中的微晶蜡。蜡中的烃类具有不同的化学结构,所以蜡的熔点与其沸点没有直接的关系。蜡是根据其熔点和油含量划分级别的。石蜡是一种含有直链烃类的固体结晶混合物,其碳链长度通常为 20 ~ 30 甚至更高。蜡组分在常温(25℃;77℉)下是固体,而石蜡油(凡士林)中既有固体烃类也有液体烃类。苏格兰地区最先使用结晶蜡脱油技术来生产石蜡,利用页岩油中蜡的不同熔点来分离蜡馏分。虽然结晶蜡脱油技术依然有所应用,但是这一工艺正在被更简单的蜡再结晶工艺逐步取代。在结晶蜡脱油中,混油石蜡,也就是粗蜡或原料蜡缓慢加热到一定温度,蜡和较低熔点的蜡中的油会变成液体,从结块底部滴下,留下高熔点的蜡残渣。

当用蜡提纯其他产品时,炼油厂要根据脱蜡工艺的分类采用一些工序;然而这些工序也一定要按照蜡生产工艺来分类。大部分工业脱蜡工艺采用溶剂稀释:通过冷却使蜡结晶,然后过滤。甲乙酮(甲乙酮—苯溶剂)工艺得到了广泛的应用。进料通过刮面式冷却器的外壁冷却形成蜡结晶,蜡和由此产生的蜡溶浆通过全封闭旋转式真空过滤机实现分离。

另一方面,中间石蜡馏分包括石蜡和性质介于石蜡与微晶蜡之间的蜡。因此,溶剂脱蜡过程生产三种不同的含油蜡,这取决于过程中加工的是轻质、中质还是重质石蜡馏分。重润滑油馏分中制得的含油蜡会以深色粗蜡出售,由中间润滑油馏分制得的蜡以浅色粗蜡出售。后者要用碱水、黏土进行处理以除去气味,提高亮度。

在丙烷脱蜡工艺中,使用旋转式压力过滤器降低压力,使部分丙烷稀释剂得以蒸发,这样浆液就可以达到合适的过滤温度。复杂脱蜡不需要冷却,但要形成固体尿素—石蜡复合物,该复合物需过滤分离,然后分解。该工艺用于制造低黏度润滑油(冷却器、变压器、液压油),必须保证产品低温下的流动性。

催化脱蜡工艺(对正烷烃进行选择性加氢裂化)使用分子筛催化剂,以保证只有烷烃分子发生有效的加氢裂化。催化脱蜡是一种加氢裂化过程,在高温($280 \sim 400$℃;$536 \sim 752$℉)和高压 $2070 \sim 3450$kPa($300 \sim 500$psi)下进行。然而具体的脱蜡操作条件取决于进料的性质和产品的倾点。该过程使用的是丝光沸石型催化剂,该催化剂具有合适的孔结构,能够选择性进行正烷烃的裂化。催化剂上的铂对活泼中间体有加氢作用,使烷烃进一步裂化仅限于初级热反应过程中。

另一种催化脱蜡过程对正烷烃也具有选择性,那些烷烃可能有较小的支链。在该过程中,催化剂经过相对温和的非氧化处理后可以恢复活性。恢复活性所需要的时间取决于进料的性质。多次恢复活性之后,催化剂上可能会结焦。催化脱蜡工艺适用于各种润滑油基础油,因此,该工艺很有可能取代溶剂脱蜡,甚至可以与溶剂脱蜡联合使用,成为减少溶剂脱蜡设备故障的一种方法。

沥青质

生产沥青质实质上是将原油中的所有可以蒸馏出来的成分都蒸馏出来,最后获得理想的残渣。该过程是分阶段完成的:原油经过常压蒸馏可以除去较低沸点的组分,得到的拔头原油可能含有高沸点(润滑)油、沥青质甚至是蜡。拔头原油通过减压蒸馏,除去一些易挥发的组分,留在塔底的产物为沥青质。该阶段的沥青质通常(错误地)被认为是沥青。为了满足产品规格,润滑油经溶剂处理后添加到硬沥青中,可使沥青变软。同时,软沥青可以通过氧化(鼓风)转化为硬沥青。

稀释沥青是用轻质油将硬沥青稀释后获得的混合物,这可使沥青质不用剧烈加热就成为液体,并得以利用。稀释液的挥发性决定着挥发速度和硬化速度,所以根据稀释液的挥发性,可将稀释沥青分为快凝、中凝、慢凝稀释沥青。沥青质不经加热就可加水乳化使用。这样的乳化通常是水包油型。当沥青质在石头或者泥土表面翻转、摩擦时,油滴就会黏附在石头上,而水会消失。该沥青除了用于铺路和固定沙土外,还用于造纸和防水。乳化剂主要分为皂型、碱型、中性与黏土型。前者触碰后易破裂,后者比较稳定,主要是靠蒸发散失大部分水分。好的乳化剂在储存或冷却时能够保持稳定性,有适当的流动性,并且易于控制破乳的速度。

近年来,沥青质逐渐成为富有价值的炼制品。20世纪80年代后期,市场上缺乏高质量的沥青质,因为当时不知道液态燃料(如汽油)对化工厂来说是多么的重要。因此曾经用来生产沥青质的残渣到现在才用来生产液体燃料(和焦炭)。

石油焦

石油焦是将渣油进行干馏(焦化)处理之后留下的残渣。石油焦的组成取决于原油的来源,但是通常来说大部分是高相对分子质量的复杂烃类(多碳少氢)。曾经有报道称石油焦在二硫化碳中的溶解度能够达到50%~80%,但事实上却不是这样,因为石油焦是不可溶的、蜂窝型的物质,它是热加工过程的最终产品。石油焦有很多应用,主要用在精炼铝所需碳电极的生产中,这需要高纯度、低灰分、无硫的碳。另外,石油焦也用来生产碳刷、碳化硅磨料、结构性碳(如管式和环式),以及生产乙炔的碳化钙。质次原油产出的石油焦为混有碳的溶剂,可作燃料燃烧,此时需要妥善处理烟道气。石油焦可用于流化床燃烧器或者发电用的燃气发生器。

Key to Exercises

Ⅰ.1. Wax production by wax sweating was originally used in Scotland to separate wax fractions by employing various melting points from the wax obtained from shale oils.

2. The catalyst employed for the process is a mordenite – type catalyst that has the correct pore structure to be selective for normal paraffin cracking. Platinum on the catalyst serves to hydrogenate the reactive intermediates so that further paraffin degradation is limited to the initial thermal reactions.

3. Asphalt manufacture is, in essence, a matter of distilling everything possible from crude petroleum until a residue with the desired properties is obtained.

4. This is usually done by stages; crude distillation at atmospheric pressure removes the lower boiling fractions and yields a reduced crude that may contain higher boiling (lubricating) oils, asphalt, and even wax. Distillation of the reduced crude under vacuum removes the oils (and wax) as volatile overhead products and the asphalt remains as a bottom(or residual) product.

5. No, it's not. The solubility of coke in carbon disulfide has been reported to be as high as $50\% \sim 80\%$, but this is, in fact, a misnomer, since the coke is an insoluble, honeycomb – type material that is the end product of thermal processes.

II. 1 – 5 ABABB 6 – 10 BAAAC

III. 1 – 5 BACAA

IV. 1. 蜡中的烃类具有不同的化学结构,所以蜡的熔点与其沸点没有直接的关系。蜡是根据其熔点和油含量划分级别的。

2. 另一方面,中间石蜡馏分包括石蜡和性质介于石蜡与微晶蜡之间的蜡。

3. 另一种催化脱蜡过程对正烷烃也具有选择性,那些烷烃可能有较小的支链。

4. 稀释沥青是用轻质油将硬沥青质稀释后获得的混合物,这可使沥青质不用剧烈加热就成为液体并得以利用。

5. 石油焦是将渣油进行干馏(焦化)处理之后留下的残渣。

V. 1. 当用蜡提纯其他产品时,炼油厂要根据脱蜡工艺的分类采用一些工序;然而这些工序也一定要按照蜡生产工艺来分类。大部分工业脱蜡工艺采用溶剂稀释:通过冷却使蜡结晶,然后过滤。甲乙酮(甲乙酮—苯溶剂)工艺得到了广泛的应用。进料通过刮面式冷却器的外壁冷却形成蜡结晶,蜡和由此产生的蜡溶浆通过全封闭旋转式真空过滤机实现分离。

2. 近年来,沥青质逐渐成为富有价值的炼制品。20 世纪 80 年代后期,市场上就缺乏高质量的沥青质,因为当时不知道液态燃料(如汽油)对化工厂来说是多么的重要。因此,曾经用来生产沥青质的残渣到现在才用来生产液体燃料(和焦炭)。

5.3 石油化工产品

导语

石油化工产品一般指直接或间接地从石油中提炼的化合物,这些化合物往往是石油炼制过程中产生的副产品。石化产品分类是为了表明化合物的来源,其名称只是用来识别原料。从石油生产化学品是基于各种类型的化合物对各种基本的化学作用感应迅速,如氧化作用、卤化作用、硝化作用、脱氢作用、聚合作用和烷基化作用。迄今为止,人们最感兴趣的是从天然气和炼厂气中所得到的低相对分子质量石蜡和烯烃以及简单芳烃,因为这些物质能够进行迅速分离和处理。

课文

石油化工产品

石油化工业始于 20 世纪 20 年代,当时炼油工艺已取得改进,可获得多种相应副产品。由于炼油工业提供了丰富而廉价的原材料,石化行业与炼油工业同步发展,并在 20 世纪 40 年代迅速发展。石油化工产品是指所有从原油(和天然气)中加工得到的化学制品(有别于燃料和石油产品),这些产品已广泛应用于商业。但广义的概念还包括脂肪类、芳香类和环烷类化学制品,还包括炭黑和诸如硫黄与氨气这样的无机物。原油和天然气都是由碳氢化合物分子组成的,碳氢化合物分子含有一个或多个碳原子,碳原子上附有氢原子。目前,石油和天然气是石油化工原材料的主要来源,因其价格低,供应稳定,生产工艺流程简单。主要的石油化工制品包括烯烃(乙烯、丙烯、丁二烯)、芳烃(苯、甲苯和对二甲苯异构体)和甲醇。因此,石油化工原料一般分为三类:烯烃、芳烃和甲醇;第四类包括无机组分和混合气体(氢气和一氧化碳的混合物)。在许多情况下,一些特殊的石化产品来源于煤、焦或植物等。例如,苯和环烷类可以从原油或者煤中制得,而乙醇可以是石油化工或者植物产品。

石油化工制品通常分为三类:(1)脂肪族类化合物,如丁烷和丁烯;(2)脂环族类化合物(如环己烷、环己烷衍生物)以及芳香类化合物(如苯、甲苯、二甲苯和萘);(3)无机物类,如硫黄、氨气、硫酸铵、硝酸铵以及硝酸。

脂肪族类

甲烷来源于原油或天然气,或是多种转化(裂化)过程的产物,是生产脂肪族类石油化工产品原材料的重要来源。乙烷也可从天然气和裂化过程中获得,是生产乙烯的主要来源。而乙烯又可生产更多有价值的石油化工产品。

乙烯是一种重要的烯烃,通常由裂化气(如乙烷、丙烷、丁烷,或炼厂废气,即这些气体的混合物)制得。当乙烯原料稀少昂贵时,就会在特殊设计的乙烯裂解炉里用石脑油甚至全部原油生产乙烯。重质原料会导致产品中的烯烃和芳烃的相对分子质量变大。在脂肪族类石油化工产品的生产中,消耗乙烯的量比其他烃类都要多,但是乙烯绝不是脂肪族类石化产品的唯一来源。丙烷和丁烷也是很重要的脂肪族类碳氢化合物。丙烷经常通过热裂化转化成丙烯,而丙烯也可以从炼厂气中得到,丁烯更常来源于炼厂气。众所周知,丁烯可以由丁烷脱氢得到,但是该方法相对于乙烯或丙烯裂化要复杂得多,其产品销售也不尽如人意。

汽油及其他液体燃料的生产需要消耗大量丁烷。炼厂产的气体成分中包含许多化学中间体,可以用来生产多种产品。合成气体(一氧化碳和氢气)也可以用于生产有价值的化工产品。

脂环族类和芳香类

环类化合物(环己烷和苯)也是石化产品的重要来源。芳香类化合物在催化重整产品中占有很大比例。芳香类化合物用于生产石化产品时要使用溶剂,如乙二醇(尤迪克斯法萃取工艺过程)和环丁砜,将其从重整产品中萃取出来。

混合的单环芳烃称为 BTX,是苯、甲苯和二甲苯的缩写。苯和甲苯通过蒸馏分离,二甲苯的同分异构体通过超精馏、分步结晶或吸附作用进行分离。苯是生产苯乙烯、苯酚、纤维和塑料的重要原料。苯和环己烷可以生产诸如尼龙和聚酯纤维、聚苯乙烯、环氧树脂、酚醛树脂和聚氨酯等产品。

甲苯可以用于制造多种化学制品,但是大部分用于调和汽油。二甲苯的用途取决于同分异构体:对二甲苯生产聚酯,邻二甲苯生产邻苯二甲酸酐。二者都可用于生产多种消费品。

大部分的苯、甲苯、二甲苯来自于石脑油的催化重整。作为粗混合物,这些芳烃是调和高辛烷值汽油的重要组分。然而目前有许多化合物由这些芳香类化合物衍生而来,因此,目前已经有一些工艺将石脑油或者重整粗柴油中的芳香类化合物有选择地提取出来,用于化工产品的生产。

无机物

氨气是目前最常见的一种无机化合物,可由氢气和氮气反应直接生成。氮气来源于空气;氢气主要来源于炼厂气、天然气(甲烷)和石脑油的蒸气转化,以及部分氧化烷烃或大相对分子质量炼厂残渣(残渣、沥青)。氨气主要用于生产硝酸铵(NH_4NO_3)以及其他铵盐和尿素(H_2NCONH_2),这些都是化肥的主要成分。

炭黑是由碳类物质(有机物)与一定量的空气进行不完全燃烧的产物,也被划分为一种无机石化产品。碳类物质生产原料广泛,甲烷、芳烃及煤焦油副产品都可以生产。炭黑主要用于生产合成橡胶。硫黄,另一种无机物石化产品,由硫化氢氧化得到:$2H_2S + O_2 \longrightarrow 2H_2O + 2S$。

硫化氢是天然气及大部分炼厂气的组分,尤其是那些加氢脱硫过程中产生的废气。大部分硫转化为硫酸再生产肥料和其他化工产品。硫还可用于二硫化碳、精制硫黄、纸浆和纸业的生产。

Key to Exercises

I . 1. The definition of petrochemicals, however, has been broadened to include the whole range of aliphatic, aromatic, and naphthenic organic chemicals, as well as carbon black and such inorganic materials as sulfur and ammonia.

2. Petrochemicals are generally divided into three groups: (1) aliphatics, such as butane and butene; (2) cycloaliphatics, such as cyclohexane, cyclohexane derivatives, and aromatics (eg, benzene, toluene, xylene, and naphthalene); and (3) inorganics, such as sulfur, ammonia, ammonium sulfate, ammonium nitrate, and nitric acid.

3. Methane, obtained from crude oil or natural gas, or as a product from various conversion (cracking) processes, is an important source of raw materials for aliphatic petrochemicals. Ethane, also available from natural gas and cracking processes, is an important source of ethylene, which, in turn, provides more valuable routes to petrochemical products.

4. Xylene use depends on the isomer: p – xylene goes into polyester and o – xylene into phthalic anhydride.

5. Carbon black, also classed as an inorganic petrochemical, is made predominantly by the partial combustion of carbonaceous (organic) material in a limited supply of air. Carbonaceous sources vary from methane to aromatic pe-

troleum oils to coal tar by – products. Carbon black is used primarily for the production of synthetic rubber. Sulfur, another inorganic petrochemical, is obtained by the oxidation of hydrogen sulfide: $2H_2S + O_2 = 2H_2O + 2S$.

II. 1 – 5　CCAAB　6 – 10　AACAD

III. 1 – 5　CABCC

IV. 1. 石油化工产品是指所有从原油(和天然气)中加工得到的化学制品(有别于燃料和石油产品),这些产品已广泛应用于商业。

2. 主要的石油化工制品包括烯烃(乙烯、丙烯、丁二烯)、芳烃(苯、甲苯和对二甲苯异构体)和甲醇。

3. 当芳香类化合物用于生产石化产品时,要使用溶剂,如乙二醇(尤迪克斯法萃取工艺过程)和环丁砜,将其从重整产品中萃取出来。

4. 氨气是目前最常见的一种无机化合物。

5. 氨气可由氢气和氮气反应直接生成;氮气来源于空气。

V. 1. 石油化工制品通常分为三类:(1)脂肪族类化合物,如丁烷和丁烯;(2)脂环族类化合物,如环己烷、环己烷衍生物,以及芳香类化合物(如苯、甲苯、二甲苯和萘);(3)无机物类,如硫黄、氨气、硫酸铵、硝酸铵以及硝酸。

2. 硫化氢是天然气及大部分炼厂气的组分,尤其是那些加氢脱硫过程中产生的废气。大部分硫转化为硫酸,再生产肥料和其他化工产品。硫还可用于二硫化碳、精制硫黄、纸浆和纸业的生产。

5.4　合成有机化学品

导语

许多物质正用来代替某些天然材料,其原因或许是人们难以获得足够的天然材料,或者是人们看中了一些合成物质的物理性能,想使之能在一定的应用领域中达到最大限度的功效。在很多情况下,合成产品的制造是有意仿制某些稀有的天然材料而产生的。此外,还必须指出,是分子的内部结构赋予合成产品特定的属性,因此这些产品的制造才会具有天然材料的全部优点,而不带天然材料的一般缺陷。

课文

在全球合成有机化工产品行业,产品和加工工艺在不断地发展变革。通过技术之间的相互连接,这些产品和工艺逐渐渗透整个产业经济,触及并使无数行业发生转变,甚至影响到了那些明显与化工生产不相关的行业。

有机合成化工产品行业为许多工业生产提供原料,包括塑料、纤维、溶剂、生化药剂、食用化学品及建筑材料等生产领域。还有一些行业越来越多地受到有机化学品的影响,包括在 21 世纪初期发展最快的关键行业,如电子通信、汽车运输、生物技术、农业、环境治理以及食品科学等。

合成有机化学品的早期发展

最早的有机化学品是取自动物或植物的天然材料,包括染织品、草药、松脂制品和溶剂、纸质品等纤维素产品,以及各种商用树脂和涂料。现代合成有机化工产品工艺的起源可追溯到 19 世纪中期,当时英国的威廉·帕琴(Willian Perkin)首次实验合成了染料。

到 19 世纪 60 年代,利用煤焦油馏分生产有机化工产品已成为化工业的重要分支。煤焦油残渣通过窄馏分切割,可以获得有机合成所需的基本芳环原料,如苯、甲苯以及二甲苯。这些物质用作合成染料、药物、树脂和溶剂的中间产品。冶金行业中研制的改良焦化炉(提高了焦化过程中煤焦油馏分的产出量)促进了合成有机化工产品业的发展。19 世纪后 25 年,钢铁工业将这类日渐增多的副产品出售给当地有机化工产品生产商。

从这段时期开始直到 20 世纪前初叶,德国取代英国成为世界上主要的煤焦油基有机化学品生产国。德国主要从其日渐发展的钢铁行业获得原料。有机化工产品行业在德国能够得到发展的另外一个重要原因在于,它能够大量生产高纯度的硫酸,而硫酸能够将煤焦油中间产品转变成最终产品。因此,在 19 世纪 90 年代,德国巴斯夫公司(Badische)发明了以催化剂接触法制成高级别硫酸的新方法,标志着该国有机行业取得了非常关键的发展。

到 19 世纪 90 年代,德国化工业几乎控制了世界煤焦油化学品市场的四分之三。主要有机化学品公司包括拜耳公司(Bayer)、巴斯夫公司(Badische)以及赫斯特公司(Hoechst)。20 世纪 20 年代,这几家公司和德国另外几家主要化工产品公司合并,最终成为世界最大的有机化学品公司——法本化学工业股份公司(IG. Farbenindustrie)。第一次世界大战之前,德国化工业为全世界提供了近 95% 的合成有机产品。1914 年,德国向美国出口的产品中仅煤焦油制得的染料就比美国染料行业总产值 2.5 倍还多。

石油与有机化学品

直到第二次世界大战后期,煤焦油馏分一直是欧洲的主要有机原料。即使在美国,在 20 世纪 50 年代之前,煤焦油一直主导一些主要化工产品生产商的生产,最著名的是杜邦公司(DuPont)。

然而在 1914 年以后,美国有机产品的主要发展趋势发生了转变,石油和天然气取代煤焦油,成为合成有机化工产品的基本原料,进而导致美国的有机产品越来越重要,可以与德国相抗衡。

在 20 世纪,美国几乎一直是主要的石油生产国。在过去的 60 年间,石油生产对美国石油化学工业的形成和发展都至关重要。

19 世纪 60 年代到 19 世纪 80 年代中期,美国国内大多数原油产自宾夕法尼亚州,更确切的产地为匹兹堡地区。1885 年以后,原油生产转移到美国中西部,尤其是俄亥俄州和印第安纳州。第一次世界大战之后,石油产业发展到俄克拉荷马州和路易斯安那州中部地区,以及新墨西哥、科罗拉多、怀俄明州和加利福尼亚州。

从 1920 年到 1941 年,石油生产发展最快的地区是墨西哥湾,该地区占美国石油生产的总量从 6% 增长到 16% 。到 20 世纪 60 年代,得克萨斯州成为最大的石油生产州,其次是加利福尼亚州、俄克拉荷马州和路易斯安那州。到 20 世纪 80 年代,墨西哥湾控制了美国石油生产的 80% 。

从结构上看,石油由烃分子组成,其中含有各种碳氢结构。碳原子的数量和排列结构决定这些碳氢化合物分子的结构。碳氢化合物分子结构可能是直链结构、带支链结构或环状结构。就相对分子质量而言,较小的分子(碳原子数较少)通常为气体;居中的通常为液体;较大的(碳原子数较多)通常为黏稠液体或固体。

碳氢化合物分为很多种类,每一类都拥有独特的分子结构与相应的化学与物理性质,因此,长久以来每类碳氢化合物都有不同的市场应用领域。原油主要由 4 种烃族组成:烷烃、烯烃、环烷烃和芳香烃。

除了以上种类的碳氢化合物,原油中还包含许多其他元素,如硫、氮、氧、金属和矿物盐。原油组成并不完全相同,会因发现地的地理位置有所变化。

美国石化行业专业从事工业用烯烃和脂肪族的生产。20 世纪 20 年代,联合碳化物公司首先获得商用烯烃乙烯,该产品为最先获得的有机中间品,且到如今还是最重要的有机中间品。继乙烯的大量生产和广泛利用之后,丙烯和丁烯也得到应用。例如,将丁烯转变成酮及其衍生物,并作为溶剂在汽车行业和其他行业得到广泛应用。

这些烯烃是美国石化行业的基础。从 1926 年到 1939 年,以乙烯为原料合成的工业化合物数量从 5 种增加到 41 种。同一时期,由丙烯为原料合成的工业化合物数量增长了近 10 倍,从 7 种增加到 68 种。

到 20 世纪 40 年代和 50 年代,美国石油化学产品开发了其他化合物族群,包括二烯烃、乙炔、石蜡,最后是芳香化合物,这些化合物最终成功取代了煤焦油基原料。直到1950年,石油化学产品的发展几乎仅限于美国。20世纪50年代,基于美国工艺设计的石油化学品生产开始扩展到欧洲和亚洲。从1960年开始,石油化学工业开始成为全球性产业。

20世纪石油炼制的产生,与美国有机化学工业的发展及其最终形成的世界影响力有着密不可分的联系。

Key to Exercises

Ⅰ.1. The synthetic organic chemicals industry supplies materials to a wide range of industrial operations. These include plastics, fibers, and solvents, bio – chemical agents, food chemicals, and materials for construction.

2. Coal tar residues, when carefully distilled, supply basic aromatic building blocks for organic synthesis: benzene, toluene, and the xylenes.

3. Germany took over from England as the world's dominant coal – tar – based organic chemical producing country. Germany obtained its raw materials mainly from its growing iron and steel industry. Also important to the future growth of the industry was Germany's ability to manufacture large volumes of high purity sulfuric acid, which was needed convert the coal tar intermediates into final products.

4. The prevailing trend for U. S. organics production after 1914 was the substitution of petroleum and natural gas for coal tars as the basis for synthetic organic chemical manufacture. This in turn resulted in the rising importance of U. S. organics vis – a – vis Germany.

5. Between the 1860s and the mid – 1880s within the United States, most of the petroleum crude came from Pennsylvania, and more particularly, the Pittsburgh area.

Ⅱ.1 – 5　CCCBD　6 – 10　ABADD

Ⅲ.1 – 5　BDBBD

Ⅳ.1. 在全球合成有机化工产品行业,产品和加工工艺在不断地发展变革。

2. 从1926年到1939年,以乙烯为原料合成的工业化合物数量从5种增加到41种。

3. 直到1950年,石油化学产品的发展几乎仅限于美国。

4. 原油组成并不完全相同,会因发现地的地理位置有所变化。

5. 20 世纪石油炼制的产生与美国有机化学工业的发展及其最终形成的世界影响力有着密不可分的联系。

V.1. 在全球合成有机化工产品行业,产品和加工工艺在不断地发展变革。通过技术之间的相互连接,这些产品和工艺逐渐渗透整个产业经济,触及并使无数行业发生转变,甚至影响到了那些明显与化工生产不相关的行业。

2. 从这段时期开始直到 20 世纪前初叶,德国取代英国成为世界上主要的煤焦油基有机化学品生产国。德国主要从其日渐发展的钢铁行业获得原料。有机化工产品行业在德国能够得到发展的另外一个重要原因在于,它能够大量生产高纯度的硫酸,而硫酸能够将煤焦油中间产品转变成最终产品。

第6章　石油炼制与环境保护

6.1　炼制污染(1)

🔲 导语

石油炼制业的工业排放和废弃物对环境的总体影响很大。在石油炼制过程中,大气会受到硫化氢、二氧化硫、氧化氮、一氧化碳、碳氢化合物以及其他有毒物质的污染。炼厂用淡水进行产品冷却,结果这些冷却水带着原油、石油产品以及矿物盐水等污染物返回到原来的水源。空气和水污染的程度取决于石油炼制所使用的技术和控制方法,同时也取决于工厂规模。

🔲 课文

处理原油及其制品的过程中,环境问题不断显现,这是由于石油工业不同于其他工业,要对世界各地的原油进行勘探、运输、精炼及储存。燃料的运输和消费就更分散了。油田和原油主要炼制与消费区不在一起,因而需要长距离的运输。这就意味着石油工业需要在各个方面采取预防措施,其中涉及大气污染、水污染、土壤污染以及破坏程度相对较轻的噪声污染等方面。由于人口密度、环境污染程度以及工业区的集中程度不同,各国制定的法律标准也会不同,因此缺乏统一的标准。

由于环保技术和措施花费很高,统一立法是绝对必要的。这一经费占整个炼制成本的1/4以上,因而必须保证国家之间的公平竞争。

以下将分别讨论原油炼制、储存、装卸及其产品使用过程中引发的环境问题,因为各个环节所存在的问题不尽相同。

生产排放

尽管炼厂采用了闭路气密性炼化装置,并在原油炼制及产品储存过程中小心谨慎,但仍会有污染物排入空气和水中,这一点难以完全避免。这取决于炼制过程中的管理控制及产品性质。

烃类化合物具有较高饱和蒸气压及一定的水溶性,因此会有一部分逸散到空气中,少量随污水排出。致癌性芳香烃类则更加危险。

更值得注意的是原油中含硫和氮的杂环化合物,不但具有臭味和毒性,在炼制时还会排放大气污染物 SO_2 和 NO_x。

在生产车间正常运转时也会有烃类泄漏,原因是:管道系统中法兰泄漏;阀门、泵以及压缩机封口泄漏;采样过程中的泄漏。在发生事故时,应将烃气体通过闭路系统引向火炬,尽量利用回收系统回收气体,压缩后返回炼制过程。剩余的气体在高架或水平火炬中燃烧,其燃烧效率可达99%。液体产品可收入装有压力水库和蓄水池的封闭积液系统中,之后回到生产流程中。

许多新工厂建设之初就采用了一些减排方法,而对现有工厂必须进行持续改造。例如,采用无法兰阀管道、低排放填料箱以及诸如多层滑盖填料环的密封件。对于易产生恶臭、剧毒及致癌物质的产品,须采用更为有效的措施(如固定电机泵、特殊萃取装置等)。

生产过程中发生的烃类排放大多发生在储存区,如原油罐区、原料储库、中间品和终端产品储库等。通常使用真空压力阀门以及浮顶罐来减少泄漏。最近在零泄漏油库方面有了较多发展,即几个装有内浮顶的固定顶储罐连入一个闭路系统,共用一个气量计,而气量计通常可以平衡所有油罐的液位。储油系统很少发生重大变化(如大量进出油、太阳辐射、暴雨等),万一发生大变化,剩余油气会在附带的火炬中进行无害燃烧,抑或在油罐压力下降时填充惰性气体补充。

大量碳氢化合物也会在装载设备中散溢,尤其是装汽油时。于是,低泄漏或零泄漏的公路、铁路及船舶运输方式广为采用。

具体可采用多种方法:

(1)油蒸气回收,即在封闭系统中泄漏的汽油蒸气可回收到油罐或者气量计中,用作炼厂的动力燃料。

(2)用合适的吸附剂对油蒸气进行再生吸附。

(3)冷却或净化蒸气后回收液态产品。

在油罐车和临近的管路及装置中,必须防止产生爆炸性汽油—空气混合物。可通过以下方式预防:将浓度控制在爆炸范围之外;缩短运输距离,排除火源;严格控制含氧量。从油罐车向加油站输油,油罐车和加油站储罐之间设置了越来越多的蒸气回收装置。

上述油品生产、储存以及运输过程中减少大气污染的措施使得西欧20世纪80年代中期的排放量减少,不到人为烃类排放总量的8%。炼厂自身的排放量仅占到1/4。

废水中的烃类物质

由于原油本身含水及在众多生产环节中使用水蒸气,在原油炼制中不可避免地会产生含烃污水。常见炼厂产生的含烃污水量为$60 \sim 100$ 米^3/时。

油水分离后,污水必须从加工装置中除去,并通过闭路系统引至净化系统。露天炼厂和油罐区的雨水以及由于事故或泄漏可能污染的冷却水也需要采用同样方法进行处理。

污水净化处理步骤如下:机械分离(筛网、过滤器、油水分离器);物理化学净化(分离、絮凝、浮选);生物净化。

生物法处理炼厂含烃废水通常没什么问题。然而必须持续监测污水流入量及相应的缓冲罐容,以便监测含硫化合物、含氮化合物与含氧化合物(如酚类)污染情况。

在许多国家,为了除臭和完全除氮进而保护地表水(河流、湖泊等),对地下水处理厂的法律要求日益增多。除氮通常要求增加额外的净化步骤,需要更长的沉淀时间。经生物净化后,水变得清洁,成为可以接受的水。

Key to Exercises

Ⅰ. 1. Because the petroleum industry — like hardly any other branch of industry — maintains exploration, transport, and refining installations for crude oil that are scattered over the entire globe.

2. Because of the high cost of technical measures for environmental conservation, harmonization of legislation is absolutely necessary.

3. Further attention must be paid to the sulfur and nitrogen compounds originating from the heteroatomic compounds in the crude oil, both because of their smell and toxicity and because of the air pollution which arises in the form of SO_2 and NO_x emissions during firing in process plants.

4. Care must be taken to prevent the formation of explosive gasoline-air mixtures. This can be achieved by (1) keeping the concentrations outside the explosive range, (2) short transportation paths and exclusion of ignition sources, and (3) extremely strict control of the oxygen contents.

5. Treatment in the wastewater purification system is carried out stepwise by:

(1) mechanical separation (sieves, filters, oil – water separators).

(2) physicochemical purification (stripping, flocculation, flotation).

(3) biological treatment.

Ⅱ. 1 – 5　ABCAA　6 – 10　BDB

Ⅲ. 1 – 5　ABDCB

Ⅳ. 1. 燃料的运输和消费就更分散了。

2. 尽管炼油厂采用了闭路气密性炼化装置,并在原油炼制及产品储存过程中小心谨慎,仍会有污染物排入空气和水中,这一点难以完全避免。

3. 更值得注意的是原油中含硫和氮的杂环化合物,不但具有臭味和毒性,在炼制时还会排放大气污染物 SO_2 和 NO_x。

4. 大量碳氢化合物也会在装载设备中散溢,尤其是装汽油时。

5. 由于原油本身含水及在众多生产环节中使用水蒸气,原油炼制时不可避免地会产生含烃污水。

V.1. 以下将分别讨论原油炼制、储存、装卸及其产品使用过程中引发的环境问题,因为各个环节存在的问题不尽相同。

2. 油水分离后污水必须从加工装置中除去,并通过闭路系统引至净化系统。露天炼油厂和油罐区的雨水,以及由于事故或泄漏可能污染的冷却水,也需要采用同样方法进行处理。

6.2 炼制污染(2)

🔲 导语

将来石油炼制能力提高,如果应对污染排放的方法仍维持在目前这个水平,那就完全不足以防止空气、水和噪声的进一步污染。因此要立即采取行动,制定一系列措施,在资本投资允许的范围内,用以不仅相对地而且是绝对地降低炼厂工业排放对环境的影响。这些措施包括引进新的炼制工艺、改良设备和装置以及组织生产的先进方法。

🔲 课文

由于烃类可溶于水(即使溶解度很小),对原油及其产品进行加工处理时,必须小心避免烃类渗透到土壤中,污染地下水。

原油从联合站输送到炼油厂过程中,大部分通过地下管道运输,这是最为安全的方式。为确保输油安全,管道采用高级钢材,阴极防腐保护,绝缘性能好,并且在高空或地面对泄漏进行连续观测。在地形复杂以及温差大的地区,需额外增加其他保护措施(管道补偿、高架管道、中间罐等)。

由于炼油厂可能污染附近饮用水源,所以选址应谨慎。根据新法律规定,所有烃处理装置必须竖直安装,防止产品溢出排放到地下、邻近街道或河道之中;储罐集中置放于由黏土层、塑料罩隔离层或者混凝土衬层建造的不渗透围堰区内;为防止泄漏,围堰区需要能够容纳整个储罐的存油量。

如果烃类已经污染了土壤,污染部分的土壤必须清除,防止继续污染地下水。如果污染土壤量较少,通常就在焚化厂燃烧处理;若污染量较大,特别是发生了大面积化学残留物和有害物质污染时,必须原地处理。可以根据土壤性质以及相应油的运移规律,在受污染地区附近的井中注入淡水,通过泵吸的方式净化污水。

用微生物降解石油的技术变得越来越重要,该方法可以就地实施,也可将土壤挖出进行外部处理。

在炼油厂外进行油品装载和储存作业均需遵守类似法规,但是每个国家法规不同。由于厂外有大量油库要运输燃料油,还有大量加油站,以及无数家庭需要燃油取暖,因而必须格外注意防止油品泄漏溢出和扩散。目前已广泛建造抗老化的耐腐蚀钢材油罐、内涂层钢制罐及新型玻璃纤维加固塑料罐。储罐建造也应考虑到一旦油品发生泄漏,这些油罐可以收纳所有油品(如双壁罐)。

在一些人口稠密、居住区和工业区密集的国家,烃萃取的处理操作规程也适用于含烃量低的水处理工艺流程。

含硫和含氮化合物

原油是天然产品,除了含有碳氢组分之外,还包含硫、氮、氧等杂原子化合物。在大气排放过程中,氮、氧化合物所占比例仅有百万分之几,不起主要作用,然而原油中硫化物的含量却可以高达几个百分点。不同炼油厂的油品中这些化合物的含量不同,相对分子质量增大,其含量也随之增大。

含硫化合物

硫及硫化合物会使催化剂中毒,并可能污染空气。硫化氢、硫醇和二氧化硫有刺鼻的臭味,并且在油品燃烧过程中产生二氧化硫,因此必须进行脱硫处理或者降低含硫量。

轻质成品油、液化石油气和汽油必须实现无硫化;最新法规要求柴油产品和轻质民用燃料油硫含量需大幅降低至 0.1% ~0.5%。

重质燃料油绝大多数用作大型工业炉燃料以及发电厂燃料,会排放大量 SO_2,存在严重问题。许多国家都规定燃料的最大硫含量为 1% ~2%。只有几种供应量有限的低硫原油不需要经过特别处理加工就可以达到这个标准。

硫元素一般是用加氢脱硫的方式从馏分中去除,即将化合物中的硫元素转化为硫化氢,然后在气体洗涤器中去掉硫化氢,并在下游的克劳斯反应过程中转换为硫单质,用作化工原料。

高度稀释之后,含有硫化氢的气体依旧气味很大,且硫化氢毒性很高,所以在处理硫化氢气体时必须采取预防措施。硫处理装置需要具备高密闭性:在高危区域,必须安装监测仪表和报警设备,一旦发生危险情况,自动关闭装置。

由于对产品的含硫量要求更严格,炼油厂的硫回收量大大增加。在传统的克劳斯装置中,如有两个反应器,硫化氢的转化峰值是95%;如果有三个反应器,峰值可以达到96%。若进一步降低含硫量,则需要进行后续处理。要求完全脱硫的国家(如德国,排放气体中硫化氢的含量须小于百万分之十),要达到这一标准,几乎只能通过额外增加高温燃烧过程来实现。

含氮化合物

大多数原油的氮含量都相对较低,加氢处理过程中馏分的含氮量会减少至百万分之几。残余的氮元素不会影响原油产品的使用。

氨在加氢过程中产生,可以添加到许多炼油厂的各个工艺阶段中,以控制加工过程(调节 pH 值),随后氨排入废水并在对废水的生物净化过程中得以清除。

噪声

过去,炼油厂对附近区域的噪声污染不构成主要问题。由于原油以及原油产品的可燃性,全球通用的做法是将泵和压缩机安装在一个开放的空间,而不是密闭的环境中。但当时的炼油厂规模较小,泵和压缩机组等动力装置制造的噪声也相对较小。除此之外,由于主要加工装置排布紧凑,同时又被附近低噪声油罐区的辅助设备所屏蔽,所以反应炉及其燃烧器对炼厂周围环境产生的声音辐射范围也较小。

尽管这样,问题还是出现了。在人口稠密的地区,工业区和住宅区间距不够大。大多数噪声源于生产流程异常运行时产生的爆鸣噪声。

随着大众环保意识的增强以及相应的立法更加严格,将来必须采取更多措施来减少噪声污染。由于技术进步,炼油厂会进行复杂的改建工作,诸如换成更多大转速、高能耗的汽轮压缩机以及增加配有更大反应炉的大型装置等,这也会使得降低噪声十分必要。

降低噪声的主要方法如下:

(1)用低噪声的燃烧器,在加热器和管道上增加隔音层。

(2)在驱动马达和涡轮泵/汽轮压缩机外加隔音罩,然而完全罩住这些设备也会有一些安全隐患,会给救火造成困难。除此之外,在密闭空间中发生爆炸的可能性也会加大。

(3)在控制阀上装隔音材料。

(4)用低噪声技术的高架火炬或者增加地面火炬。

在目前的工艺水平下,将噪声降低10分贝,即将检测值降低一半是可以达到的,但这需要大量开支。

Key to Exercises

Ⅰ. 1. Hydrocarbons' penetration into the soil will cause possible contamination of the groundwater.

2. Transport from the oil terminal to the refinery is carried out almost exclusively in underground pipelines, which are also the safest means of transport.

3. The location of the refinery must be carefully selected with regard to possible dangers to drinking water.

4. According to new legislation, all HC – handling units must be erected so as to prevent the discharge of spilled product to the underground, to adjacent streets, or canals; the storage tanks are placed in collection spaces which are made impermeable to oil using clay layers, plastic tilts, or concrete lining; in case of a leak they must be capable of receiving the entire tank contents.

5. (1) Low – noise burners and additional noise insulation on process heaters and piping.

(2) Sound hoods on the drive motors and turbines of pumps and compressors.

(3) Sound insulation on control valves.

Ⅱ. 1 – 5　ADBCC　6 – 10　BBDBC

Ⅲ. 1 – 5　CAABC

Ⅳ. 1. 由于厂外有大量油库要运输燃料油,还有大量加油站,以及无数家庭需要燃油取暖,因而必须格外注意防止油品泄漏溢出和扩散。

2. 不同炼油厂的油品中这些化合物的含量不同,相对分子质量增大其含量也随之增大。

3. 只有几种供应量有限的低硫原油不需要经过特别处理加工就可以达到这个标准。

4. 大多数原油的氮含量都相对较低,加氢处理过程中馏分的含氮量会减少至百万分之几。

5. 在目前的工艺水平下,将噪音降低 10 分贝,即将检测值降低一半是可达到的,但这需要大量开支。

Ⅴ. 1. 随着大众环保意识的增强以及相应的立法更加严格,将来必须采取更多措施来减少噪声污染。由于技术进步,炼油厂会进行复杂的改建工作,诸如换成更多大转速、高能耗的汽轮压缩机以及增加配有更大反应炉的大型装置等,这也使得降低噪声十分必要。

2. 氨在加氢过程中产生,可以添加到许多炼油厂的各个工艺阶段中以控制加工过程(调节 pH 值),随后氨排入废水并在对废水的生物净化过程中清除。

6.3 消费污染

🔲 导语

石油产品在炼厂外的燃烧过程比石油炼制过程本身排放的二氧化硫和氮氧化合物更多。油品从油库运到加油站也会排放碳氢化合物。机动车辆碳氢化合物的排放是由于汽油的饱和蒸气压较高,汽车在加油和行驶过程中会蒸发碳氢化合物,而且车辆尾气中有未完全燃烧的碳氢化合物。一氧化碳、氮氧化物、未燃烧的烃类以及含铅化合物伴随汽车尾气也一同排出,当然也会排出二氧化碳和水蒸气等燃烧产物。空气污染物中三分之一的碳氢化合物是由汽车排放的。

🔲 课文

与炼油过程本身的排放量相比,石油产品在炼厂外的燃烧过程排放了更多的二氧化硫和氮氧化合物。此外,油品从成品油库运到加油站也会排放碳氢化合物,而在运输领域中,汽车行驶以及反复加油使碳氢化合物的排放量更大。

运输燃料

当汽油在点燃式发动机中燃烧时,会有不同含量的一氧化碳、氮氧化物、未燃烧的烃类以及含铅化合物伴随汽车尾气一同排出。此外,二氧化碳和水蒸气等燃烧产物也会排出,其排量多少会因驾驶方式、燃油种类和发动机构造而有所不同。若为柴油车,排放物还将包括二氧化硫和煤烟颗粒。除了尾气排放,汽车加油以及运行过程中产生的烃气排放也是造成环境污染的一个重要因素。

车用汽油

机动车辆碳氢化合物的排放有几个原因:首先,车辆尾气中有未完全燃烧的碳氢化合物;其次,由于汽油的饱和蒸气压较高,汽车在加油和行驶过程中也会蒸发碳氢化合物。空气污染物中三分之一的碳氢化合物是由汽车排放的,占人为有机物排放总量的 12%。

由于汽车燃料广泛应用,人们必须特别注意苯的排放。汽油中苯含量应遵循怎样的标准($1\% \sim 5\%$仍存在争议)仍是石油工业面临的问题,特别是在无铅汽油生产中,各种调和油料中苯含量非常高,苯含量取决于原油的质量和来源,在高质量重整油中占8%,在裂解汽油中占到$18\% \sim 40\%$。鉴于苯提取成本过高以及苯再利用还存在问题,从碳氢化合物中提取苯难以实现,因此必须使苯蒸发减小到最低程度。

柴油机燃料

柴油机车尾气的主要问题是其尾气中含有煤烟颗粒和SO_2。相比火花点燃式发动机,柴油机中NO_x和CO组分并不那么重要。由于柴油机尾气具有较强烈的气味并且其尾气中含有的煤烟颗粒具有致癌作用,因此更需要注重减少柴油机尾气的排放。

燃烧用燃料

世界大部分炼制产品是用来作民用燃料、工业加热燃料和能源生产燃料的。这些炼制产品是液化石油气中间馏分和重燃料油所产生的馏分。SO_2、NO_x和CO的排放与燃烧程度有关,燃料的选择和预处理以及燃烧器的设计极大地影响SO_2、NO_x和CO排放,以及烟灰排放和未完全燃烧的碳(以煤烟形式存在)的排放。这种情况在以高硫重质燃料油为动力的大型加热装置中尤为突出。

液化石油气

经炼厂预处理后,液化石油气几乎不含硫,用作燃料在燃烧过程中也不产生煤烟,不会产生危害。

轻质燃用油(2号燃料油)

由于原油产地不同,中间馏分油硫含量可高达1.5%(质量分数)。炼厂通过加氢脱硫过程能够除去其中大部分的硫。

因为用于民用加热,大部分国家要求轻质燃用油的硫含量小于0.5%(质量分数),而欧盟则要求民用燃油的硫含量低于0.2%。欧洲石油化工学会的调查研究表明,若要实现更低的含硫标准,当前种类的原油脱硫成本将显著增加。

重质燃料油(6号燃料油)

原油炼制后的残渣可用作重质燃料油,供工业加热和发电使用。对于重质燃料油,必须考虑二氧化硫、氮氧化物、CO和微粒的排放,采取环境保护措施。

环境保护成本

由于对环境问题的认知程度、技术手段以及法律制度均在迅速发展,目前环保成本极高,大大影响了整体石油炼制成本。投资成本除了用于添置新设备(总成本的 15% ~20% 用于环境保护)和改造旧设备外,还要支付运营过程中的能源、维修和人力费用等。石油炼制中最重要的设施是:

(1)从硫化氢中提取硫黄,然后再净化的洗气系统和克劳斯装置。

(2)气态(通过火炬)和液态烃排放的闭路系统。

(3)原油和成品油浮顶储罐。

(4)挥发性成品油储存和装载过程中的碳氢化合物蒸气回收装置。

(5)生产装置中散逸烃气和沉淀水的收集区,油库、装载设备;废水净化设备中的排水系统。

(6)燃烧设备内的减排(SO_2,NO_x)装置。

(7)测量空气和废水污染物的封闭采样系统、实验室分析和在线分析设备。

以下设施用于生产污染相对较轻的油品:

(1)汽油和中间馏分的加氢脱硫装置。

(2)生产无铅汽油基础原料高辛烷值组分的重整、异构化装置。

(3)生产适合无铅汽油组分(甲基叔丁基醚)的合成装置。

(4)将重质残渣油转化成轻质清洁产品的装置。

(5)无铅运输燃料油的独立储存和装载设备。

🔲 Key to Exercises

Ⅰ. 1. When gasoline is combusted in the spark ignition (Otto) engine, varying amounts of CO, NO_x, unburnt hydrocarbons, and lead compounds are emitted with the exhaust gas in addition to the combustion products water and CO_2.

2. Different factors cause emissions of hydrocarbons from motor vehicles: firstly, the unburnt hydrocarbons in the motor vehicle exhaust, and secondly, the hydrocarbons emitted during refueling and running, because of their high vapor pressure. The proportion in the emissions from the motor vehicle sector is one – third of the total output, corresponding to 12% of the total man – made emissions of organic substances.

3. The liquefied petroleum gases used as fuels are almost sulfur – free as a result of pretreatment in the refineries, and they burn without formation of soot. Their use presents no problem for the user.

4. The permissible sulfur limits for light heating oil in most countries are

< 0. 5 % (wt),because of its use in domestic heating.

5. Because the understanding of environmental problems, technical solutions, and legislation are still in rapid development, the costs of environmental conservation have reached a high level.

Ⅱ.1 – 5　BDBAA　6 – 10　BBBBD

Ⅲ.1 – 5　ACAAC

Ⅳ.1. 除了尾气排放，汽车加油以及运行过程中产生的烃气排放也是造成环境污染的一个重要因素。

2. 空气污染物中三分之一的碳氢化合物是由汽车排放的，占人为有机物排放总量的12%。

3. 由于汽车燃料广泛应用，人们必须特别注意苯的排放。

4. 由于柴油机尾气具有较强烈的气味并且其尾气中含有的炭黑颗粒具有致癌作用，因此更需要注重减少柴油机尾气的排放。

5. 经炼厂预处理后，液化石油气几乎不含硫，用作燃料在燃烧过程中也不产生煤烟。

Ⅴ.1. 由于汽车燃料广泛应用，人们必须特别注意苯的排放。汽油中苯含量应遵循怎样的标准(1% ~5%仍存在争议)仍是石油工业面临的问题。特别是在无铅汽油生产中，各种调和油料中苯含量非常高，苯含量取决于原油的质量和来源，在高质量重整油中占8%，在裂解汽油中占到18% ~40%。鉴于苯提取成本过高以及苯再利用还存在问题，从碳氢化合物中提取苯难以实现，因此必须使苯蒸发减小到最低程度。

2. 世界大部分炼制产品是用来做民用燃料、工业加热燃料和能源生产燃料的。这些炼制产品是液化石油气中间馏分和重燃料油所产生的馏分。SO_2、NO_x 和 CO 的排放与燃烧程度有关，燃料的选择和预处理以及燃烧器的设计极大地影响 SO_2、NO_x 和 CO 排放，以及烟灰排放和未完全燃烧的碳(以煤烟形式存在)排放。这种情况在以高硫重质燃料油为动力的大型加热装置中尤为突出。

6.4　化工产品污染

导语

第一次世界大战利用无机材料制造出大量先进炸药，而第二次世界大战使用新型合成材料制造轮胎、降落伞和通信设备等。第二次世界大战后，合成橡胶和纤维广泛生产使用。20世纪90年代后期，90%以上的有机物生产都是以石油或天然气为原料。第二次世界大战后,空气污染问题开始显现,随后出台了各种石油化工产品监管措施。石油化工行业也认识到遵从环境法规有利于提高生产能力。

🔲 课文

石油化工产品的兴起

有机石油化工业于第一次世界大战结束后在美国东部兴起。之后30年间,石油和煤炭、谷物一样成为生产合成有机化学品的原料。

20世纪20年代到30年代,有机石油化工业集中发展烯烃新工艺技术,其中包括乙烯、丙烯和丁烯技术。在第二次世界大战前,石油炼制业在有机化学品生产中占有的份额很小。20年代早期,新泽西标准石油公司(埃克森石油公司)开发出大规模生产合成异丙醇的工艺。此外,20世纪30年代,壳牌公司利用炼厂废气中有机物和天然气作为原材料生产出合成氨。

早期的石化产品基本上都是化学品公司生产的。到20世纪20年代后期,联合碳化物公司的子公司碳化物和碳化工公司(Carbide and Carbon Chemicals Company)从西弗吉尼亚天然气中生产出乙烯及其衍生物。其中乙烯经直接裂解天然气和液态石油馏分制得,而丙烯和丁烯则作为乙烯生产的副产品获取,或者从炼油过程中制得。联合碳化物公司在这个时期开发的重要合成产品有环氧乙烷和乙烯基塑料,以及最早用作汽车散热系统合成防冻剂之一的乙二醇。此外,还生产出乙二醇醚产品用作汽车新表面涂层。

第二次世界大战成为石油合成有机物的分水岭。第一次世界大战利用无机材料制造出大量先进炸药,而第二次世界大战使用不同类型的材料,开辟了新的应用领域。坦克、卡车和飞机需要大量高辛烷值燃料;轮胎、降落伞、通信设备等需要用新型合成材料制作。正是在第二次世界大战期间,石油炼制业才积极将其石油加工能力转变成生产战略性关键石化产品的能力,尤其是合成橡胶和纤维。

第二次世界大战后石化产品在许多方面迅速发展。就地域而言,欧洲和亚洲大量采用石油化工工艺技术,其中许多是引用美国技术。同样,在美国许多地区石化产品生产也初具规模。到20世纪90年代后期,不仅伊利诺伊州、俄亥俄州、密歇根州和西弗吉尼亚州,而且得克萨斯州、路易斯安那州和俄克拉荷马州也都成为有机化学产品重要的生产中心。

另外,过去用煤或谷物生产的化学产品开始越来越多地利用更为经济的石油工艺生产。而且乙烯和氨等成熟石化产品开始在规模更大和技术更先进的工厂进行生产。

到 20 世纪 60 年代,全球 70% 多的合成有机产品源于石油或天然气。到 70 年代,约 3000 种化学产品大量产自化石燃料。90 年代后期,90% 以上的有机物生产都是基于石油或天然气。这些原料生产出 6000 多种商用有机化合物。在整个 90 年代,美国一直是世界主要石化产品中心。

催化剂和石油化工产品

自从 20 世纪早期,催化剂在有机合成化学品生产中一直发挥着主导作用。催化剂一般由一种或多种金属以及金属化合物组成,其主要作用是加快化学物质反应速度,使生产经济可行。

早期商用催化剂,无论是铁基还是铂基,都是德国开发的,主要是用于大规模生产无机重化学品——硫酸。20 世纪 30 年代,美国炼油公司太阳石油公司(Sun Oil)引进首个催化裂解技术,并投入商业生产。之后的十年间出现了更多的先进的催化裂解和重整技术,用于大规模生产,在 20 世纪 40 年代达到顶峰。

到 20 世纪 40 年代后期,催化剂用于生产一些重要有机合成物的中间品,诸如生产先进燃料,以及生产如乙烯基塑料、尼龙、合成橡胶和甲基丙烯酸酯等聚合物。20 世纪 60 年代后期和 70 年代早期,催化剂开始应用于环保领域,成为控制汽车尾气排放的催化转换器的核心成分。

到 20 世纪 90 年代后期,催化剂在有机物生产中的比例大大超过 60%,在目前有机化学品生产中占 90%。1999 年,催化剂生产在炼油、聚合、化工和环境治理四大领域所占市场份额超过 100 亿美元,全球生产催化剂的公司有 100 多家,其中一半以上在美国。

环境趋势

第二次世界大战结束时,石油化工业开始关注环境问题。1945 年,石化行业开始从天然气和炼厂气中的硫化氢中回收提取硫。其原因就是 20 世纪 40 年代末到 50 年代早期出现的大气污染和世界硫短缺问题。这种硫回收技术主要靠使用新颖且高效的吸附剂。

从石油生产化学品是基于各种类型的化合物对各种基本化学作用感应迅速,如氧化作用、卤化作用、硝化作用、脱氢作用、聚合作用和烷基化作用。因此,石化产品中的化学污染物在环境中的释放是不可避免的。

美国颁布有毒物质控制法案的目的就是授权美国环境保护机构对新型商用化工产品进行上市前的监管,当现有化工产品对人身健康和环境构成重大威胁时进行干预,限制这些产品的销售和使用。美国环境保护机构基于其对化工产品的调研数据,评估、规避和控制产品在生产和使用过程中可能产生的风险,从而限制商用化工产品的生产和销售,并要求对造成重大环境威胁的化工产品做出明确说明,或者进行其他限制。

从 20 世纪 70 年代开始,一些重要门类的有机化学品受到监管控制,特别是清洁空气方面尤为突出。受到监管的有机化学品包括甲基类燃料添加剂和以氯氟烃(CFCs)形式存在的卤代烃。

最近几年,由于环境监管需求越来越高,有机化学品公司投资环保技术已达数十亿美元。此外,它们还尝试将有益环境保护的技术整合到生产工艺中,把更高程度的再循环能力纳入其生产设计。在某种程度上,这也是为了防止未来严格的环保法规带来更大的生产成本。借助这些措施,有机化学品公司希望使监管机构相信化工行业能够实现自我监管。

另外,这也是出于商业考虑。像国际标准化组织(ISO)这样的机构正在使国际标准发挥越来越重要的作用,这就意味着未能达到 ISO14000 标准要求的公司将会在竞争中处于不利地位:客户将会避免和这些公司合作,而更愿意与注重环保的公司合作。化工行业也认识到遵从环境法规与提高生产力之间有着直接联系,这是有机化工企业极为关注的一点。

Key to Exercises

Ⅰ. 1. Ethylene was obtained by the direct cracking of natural gas and liquid petroleum fractions. Propylene and the butenes were derived either as a by-product of ethylene production or from refinery operations.

2. Whereas World War Ⅰ utilized the inorganics sector for making large quantities of advanced explosives, World War Ⅱ consumed different sorts of materials for new applications.

3. In the late 1960s and early 1970s, catalysts entered the environmental arena as the central component in catalytic converters for the control of automotive emissions.

4. This is important both for shipping raw materials into the plant and transporting chemical intermediates and final products out.

5. To meet rapidly growing regulatory requirements and there are also commercial consideration.

Ⅱ. 1-5 ABBAA 6-10 CACCA

Ⅲ. 1-5 ADCCC

Ⅳ. 1. 早期的石化产品基本上都是化学品公司生产的。

2. 第一次世界大战利用无机材料制造出大量先进炸药,而第二次世界大战使用不同类型的材料,开辟了新的应用领域。

3. 正是在第二次世界大战期间,石油炼制业才积极将其石油加工能力转变成生产战略性关键石化产品的能力,尤其是合成橡胶和纤维。

4. 就地域而言,欧洲和亚洲大量采用石油化工工艺技术,其中许多是引用美国技术。

5. 20世纪60年代后期和70年代早期,催化剂开始应用于环保领域,成为控制汽车尾气排放的催化转换器的核心成分。

V. 1. 联合碳化物公司在这个时期开发的重要合成产品有环氧乙烷和乙烯基塑料,以及最早用作汽车散热系统合成防冻剂之一的乙二醇。

2. 像国际标准化组织(ISO)这样的机构正在使国际标准发挥越来越重要的作用,这就意味着未能达到ISO14000标准要求的公司将会在竞争中处于不利地位。客户将会避免和这些公司合作,而更愿意与注重环保的公司合作。

References

[1] Matthew P Brouwer. Oil refining and the petroleum industry. Nova Science Publishers, 2012.

[2] William L Leffler. Petroleum refining for the non – technical person. PennWell Books, 1985.

[3] William L Leffler. 石油炼制. 北京:石油工业出版社,2012.

[4] Mohamed A Fahim, Taher A Alsahhaf, Amal Elkilani. Fundamentals of petroleum refining. Elsevier, 2010.

[5] James H Gary, Glenn E Handwerk, Mark J Kaiser. Petroleum refining: technology and economics. 5th ed. CRC Press, 2007.

[6] William L Leffler. Petroleum refining in nontechnical language. PennWell, 2000.

[7] Rathi Rakesh. Petroleum refining processes. SBS Publishers & Distributors Pvt. Ltd. ,2007.

[8] Energy Information Administration, US Department of Energy(2007). Refinery Outages: Description and Potential Impact on Petroleum Product Pricesin March,2007.

[9] Hastings, Justine, Jennifer Brown, Erin Mansur. Reformulating Competition? Gasoline Content Regulation and Wholesale Gasoline Prices. Journal of Environmental Economics and Management, January 2008.

[10] Surinder Parkash 著. 石油炼制工艺手册. 孙兆林,王海彦,赵杉林,译. 北京: 中国石化出版社,2007.